Microcontroller Cookbook

£19.99

2001

Microcontroller Cookbook

Second edition

Mike James

Newnes

OXFORD   AUCKLAND   BOSTON   JOHANNESBURG   MELBOURNE   NEW DELHI

Newnes
An imprint of Butterworth-Heinemann
Linacre House, Jordan Hill, Oxford OX2 8DP
225 Wildwood Avenue, Woburn, MA 01801–204
A division of Reed Educational and Professional Publishing Ltd

Ⓡ A member of the Reed Elsevier plc group

First published 1997
Reprinted 1998, 1999
Second edition 2001

British Library Cataloguing in Publication Data
A catalogue record for this book is available from the British Library

ISBN 0 7506 4832 5

Composition by Genesis Typesetting, Laser Quay, Rochester, Kent
Printed and bound in Great Britain by Biddles Ltd *www.biddles.co.uk*

Contents

Preface

Many students of microprocessor-based systems have some initial difficulties relating to the software and hardware. The information presented here is mostly available from all the manufacturers' data manuals, but I have tried to present it in a much more readable form and in a more logical order for the beginner. I have chosen to focus on the MCS51 and PIC families since, at the present, they seem to be more popular. Whether this is because of their in-built technical advantages or because of marketing hype is an interesting point. Maybe as your experience grows with these and competing products, you may form your own opinions on the matter. It has been an interesting task to chart the similarities and differences between the two families conceived a decade apart. The aim of writing a book which looks at two microcontrollers lies in the hope of giving confidence that all such devices have many similarities; and when you look at the differences, they are not so great either. Thus, when a technical problem has to be solved, it can be approached from the point of view of using the best tool for the job, not a case of reusing the only tool with which you are familiar.

I have not considered the problems of successful code generation and verification. In practice of course, this occupies the larger part of the development process. (If you disagree with this statement, then I would presume that is because it seems to you that more time is passed debugging non-functional code.) The aim of the book is only to look into the hardware and software structures and to give examples of how to develop short modules to test functionality.

I have not dwelt overmuch on the many varieties of software available – cross-assemblers, linkers etc. Microchip give theirs away 'free' with the PIC development systems. Other free cross-assemblers can be obtained for the MCS51 from the web sites which specialize in 8051 and related processors. Some even delve into higher level language tools such as C or Basic or Forth cross-compilers/interpreters. For professional use, there are many (costly) full-featured software and hardware development tools. Generally these improve development productivity, but are not essential for the beginner to get the full flavour of these microcontrollers. (Indeed, there are more than a few small companies which rely on these free/low-cost development tools as their main aid.) All of this software usually comes with adequate help files, and it then just falls to the programmer to get comfortable with the assemble/link process. Before committing to firmware, it is sometimes useful to use simulation software to check if the program will function correctly. Once again, software simulators are freely available which operate with varying degrees of sophistication. It is possible to completely simulate the entire project, but most developers tend to use these simulators to quickly check the functional correctness of modules or short routines.

This second edition has given me the opportunity to correct the (thankfully) few mistakes which have been pointed out. There have also been many requests to include some problems which consolidate the understanding of the preceding theoretical sections. These start reasonably blandly and progress to more sophisticated applications for which a microcontroller might be the solution.

Although the technology has moved on, with additional functionality built in to each microprocessor, I have deliberately not overextended the range of devices used. To do so would have diluted the aim of the book which is to teach the basics and compare two different microprocessor families.

Acknowledgements

I would like to thank and acknowledge the Intel, Microchip, Philips, National Semiconductor, Maxim and Texas Instrument Corporations for the use made of various data in the preparation of this book. Also to a decade-and-a-half of (mostly willing) students who have acted as guinea pigs in my efforts to present the challenging world of microcontrollers in a logical, non-baffling and interesting way.

Chapter 1

Microcomputer Systems

Introduction

A cookbook is not usually meant to be read from end to end. The reader knows what effect he or she wishes to achieve, and just browses the recipes of direct interest. So it is intended with this book. I have tried to cover all the aspects of using two of the most common and popular microcontrollers. Both of these can now be considered to be mature and well-established products which have their individual adherents and devotees. There are a lot of variants within each family to satisfy most memory/ input-output/data conversion/timing needs of the users. I have taken the embedded controller market as my target, and used as many examples of hardware and software design as I thought relevant and appropriate (not just my opinion, but from many and diverse industrial colleagues).

Histories

Microcontrollers were introduced when the skills of the semiconductor industry allowed several functions such as CPU, memory and I/O to be integrated onto one piece of silicon. This in turn reduced the size (and usually the power consumption) of any microprocessor-based solution within a problem. Because of the low cost and ease of integration within an application, they are used whenever possible to reduce the chip count of a piece of electronics. The word 'intelligent' tends to be applied to any device which contains some sort of processing/memory capability. The intelligence, of course, belongs to the hardware and software designers who program the device.

Intel introduced the 8051 family in 1980, and from a simple microcontroller, it has grown into at least 30 different versions and been second-sourced by many manufacturers. One of the difficulties of writing a book such as this is the moving nature of the target at which I am aiming. Each week brings out new, better, faster, lower-power, larger-memory versions. I have included some of the more interesting variants – usually those which easily solve a specific problem. For example, it seems somewhat perverse to implement a serial communication channel via a digital I/O port when both products have perfectly good family members with features which can handle this task automatically. Having said that, there are occasions when it is necessary in the interests of cost or compactness to solve this problem. Thus, there are examples of code structures to do this (page 46).

The newcomer PIC from Microchip Technology was developed in the late 1980s and was marketed with the two similar versions – a PROM-based version and an EPROM version. The earlier models were smaller (0.3 inch) width and had less I/O capability. However, the manufacturers came up with a cheap programmer which included assembler and simulator software. This immediately gave them a market advantage, since one of the difficulties when a designer considers a new microprocessor is the cost of the development system. It is one thing to have to learn a new structure and instruction set, without the bother of having to fight the budget holders for the cost of the hardware as well. So, for around £100, an engineer could obtain enough hardware and software to evaluate the product effectively. The product family developed, and the company now have a considerable market share.

Brief Comparisons

8051

The 8051 was introduced to replace the older 8048 family which was originally used as a keyboard handler for IBM PCs. The original product lineup was as follows:

ROM-based 8051   The program memory is on-chip and is specified by the user. The manufacturer mask-programs the silicon die so that the CPU can only execute the one program. There is a cost advantage if the IC is bought in sufficient quantities.

EPROM-based 8751   This variant is used mostly for development or very small production runs. It is quite an expensive item in one-off costs, but has the advantage that the program can be erased and a new version 'blown' into the internal memory.

ROMless 8031   This needs external memory (EPROM) to hold the program. To access the memory, the 8031 has to use some of its input/output (I/O) ports. This means that an 8031-based microcontroller has only 14 I/O lines left to directly interface to the world.

When referring to the family, most engineers simply refer to the 8051. It was introduced at a time when a single +5 V supply was a pleasant relief to digital design engineers. The original 40-pin DIL package was easily configured and the availability of 4 × 8-bit ports was pure luxury.

The basic device was blessed with:

- 128 bytes of on-chip RAM (expandable to 64 kbytes externally)

- 32 I/O lines

- two 16-bit counter timers

- 6 sources of program interrupt (each vectoring (jumping) to 6 different addresses and with priorities set by the programmer)

- Full duplex serial port (full duplex = the ability to simultaneously receive and transmit serial data).

PIC

The original 8-bit PIC microcontroller was introduced with a small number of variants with a choice of I/O ranging from the 13-line 16C54 to the 21-line 16C57. These were originally conceived as One Time Programmable (OTP) or as EPROM-based versions. An attraction of the plastic OTP version was that it was very cheap. The EPROM version, as in all microcontrollers, is relatively expensive. This is because of the expense of the ceramic body needed to be formed around the quartz UV transparent window which is used to erase the memory. The smaller package sizes were an interesting development, coming as they did in the 0.3 inch 'skinny DIP' IC. The larger I/O configurations reverted to the same 0.6 inch wide package as the 8051 for the DIL IC. All devices are available in other packages such as flatpack, leadless chip carrier, plastic leaded chip carrier (PLCC), small outline IC (SOIC), plastic shrink small outline (SSOP) and other variations on these themes. In this book, I will concentrate on Dual In Line ICs.

Like the 8051, there are now so many versions of the device, that almost any memory or interface requirement can be catered for. However, the basic PIC comes equipped with:

- small, compact instruction set (33 or 35 instructions)

- 32 bytes of RAM

- 13/21 I/O pins

- one 8-bit Real Time Clock Counter.

Written like this, the basic PIC seems to be very underequipped when compared with the basic 8051, but it's very much a case of choosing the right version. This will be dealt with in the PIC section.

Microprocessors are fairly stupid machines. They are capable of performing very simple tasks very quickly. It is this relative speed that gives them the reputation for 'intelligent' or 'smart' functions. They all have certain characteristics in common, and this section looks at these common functions before specializing in the two specific families.

Microprocessors fetch instruction codes from memory, find out what that instruction is commanding it to do, and then they do it. This is formally called the 'FETCH-EXECUTE CYCLE'. The three main areas of microprocessor-based systems are the

- **CPU**   Central Processing Unit

- **Memory**   Read Only or Read/Write

- **I/O**   Input/Output.

Instructions can cause

- Memory ↔ CPU data transfer

- I/O ↔ CPU data transfer

- Data manipulation within the CPU.

Microprocessors can be called 4-, 8-, 16-, 32- or even 64-bit machines. This characterization refers to the number of bits in a data word, i.e. the 'width' or size of data that is handled in one operation. (However, even this simple definition doesn't cover all situations: the Intel 8088 reads data in from memory in 8-bit chunks but deals with it internally in 16-bit words.) The most common term is the 'byte' or 8 bits of data. Half of this is the 'nibble'. A general-purpose term which does not need to be too specific about the size of the data is the 'word'.

Internally, all microprocessors have 'registers' which are capable of holding a data word. The number of registers depends on the type of processor. Data in a register can be transferred to another register, a location in memory, or to an input/output device. One register which is common to all microprocessors is the Program Counter. This always holds the memory address of the next instruction to be executed. Another common register is the Instruction Register. This is where each instruction is held while it is being decoded. After that, the particular action to take place depends on the microprocessor.

The Arithmetic and Logic Unit is another common element in all microprocessors. This device allows data in registers or memory to be compared or subjected to various arithmetic/logical operations such as addition, subtraction, multiplication, division, logical 'AND'ing, 'OR'ing, 'NOT'ing or 'EXCLUSIVE-OR'ing. Appendix B explains these logic operations and their various uses.

There are many software examples in this book, but not many complete programs. When trying to get the hang of a new microprocessor, I have always found that the example software almost, but never quite, solves the problem I am wrestling with; or that the author has tackled it in a different way to the way I would have gone about it. Thus, in this book, I have shown how to achieve certain effects without getting too specific about the nature of the problems. The best source of full examples is over the Internet. Sites are available at:

- 8051   http://www.cs.nmt.edu/~ketan/8051.html

- PIC   http//www.microchip.com

The amount of software at these sites is huge. There are examples of all sorts. However, I hope that with this book, you will be able to start appreciating the potential and capabilities of these devices. An internet search on '8051' or 'PIC' will turn up links to example code and assemblers/simulators and compilers from a diverse range of high level languages such as C, BASIC and FORTH. One such site is www.idiom.com/free-compilers/.

Chapter 2

The MCS51 Microcontroller

The 8051, like many microcontrollers, has the CPU, memory and I/O (Input/Output) integrated together in a flexible and extendable manner. It is an established product which is very popular in many different industries. The core features of the device are

- 8-bit CPU optimized for control applications

- Extensive Boolean processing (single-bit logic) capabilities

- 64K Program Memory address space

- 64K Data Memory address space

- 4K bytes of on-chip Program Memory

- 128 bytes of on-chip Data RAM

- 32 bidirectional and individually addressable I/O lines

- Two 16-bit timer/counters

- Full duplex UART

- 6-source/5-vector interrupt structure with two priority levels

- On-chip clock oscillator.

The basic architectural structure of the core is shown in Figure 2.1.

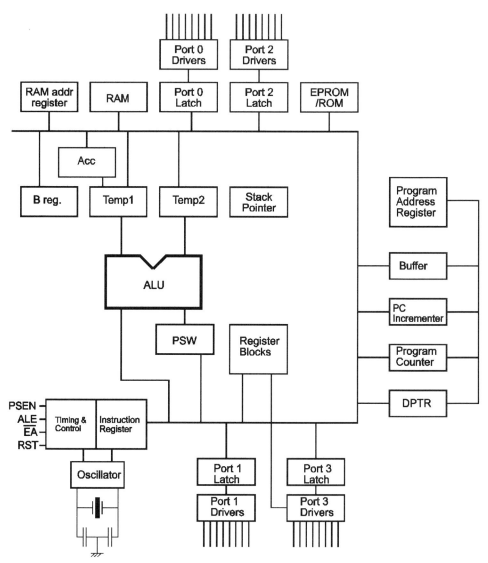

Figure 2.1 Architectural structure of the MCS51

Family Members/Variants

The MCS51 microcontroller family comes in a variety of options, and a selection of these are summarized in Table 2.1.

Table 2.1

DEVICE	EPROM	ROM	RAM	8-bit I/O	Clock	UART	ADC
8031	–	–	128	4	33	1	
8032	–	–	256	4	33	1	
8051	–	4k	128	4	33	1	
8052	–	8k	256	4	33	1	
8751	4k	–	128	4	33	1	
8752	8k	–	256	4	33	1	
80C452	–	–	256	5	16		
83C452	–	8k	256	5	16		
87C452	8k	–	256	5	16		
8×C552	8k	–	256	5	24	$1 + I^2C$	8×8
8×C652	8k	–	256	4	24	$1 + I^2C$	
87C752	2k	–	64	2 + 5/8	40	I^2C	5×8

Basically, the options available from the MCS51 family are:

- A microcontroller with internal mask-programmed ROM

  This is the option chosen for mass-produced items. The act of specifying and creating a mask-programmed ROM is an expensive capital outlay, but the unit cost falls to a very low level if the economies of scale determine that enough devices can be ordered.

- A microcontroller with internal EPROM

  This option is chosen for prototyping/developing microcontroller software, or for short production runs. The EPROM-based 8751 is relatively expensive and has the limitation that it will only tolerate 100 erase/program cycles (manufacturer's figures). Against this high(ish) initial outlay, it must be remembered that these devices hold software. The cost of developing software is the most significant part of any project.

Both the ROM- and EPROM-based versions have an inbuilt 'security fuse'. This controls the memory read mechanism, so that an external device cannot read the ROM (i.e. 'steal' the software).

● A microcontroller with external program memory

This is the cheapest option. The drawback is that some of the I/O pins now have to be used for accessing the external memory device(s). This is acceptable if the device additionally has to access other peripheral or memory components. It is also a less expensive development method since EPROM programmers are cheaper than 8751 programmers. The exact interface circuitry is detailed on page 19. The cost of the 8031 is of the same order as the basic PIC model; however, memory and interface still have to be paid for. In Figure 2.2(a), Pins 32–39, 16, 17 and several of 21–28 (depending on memory size) would be needed for external memory devices.

● 'Shrunk' microcontroller

The DS750 family is a reaction to consumer demand for a physically smaller device (0.3 inch pin pitch) with all functions integrated onto the IC. These small (24-pin DIL) microcontrollers come in EPROM or OTPROM program memory. They are not designed for external memory expansion, so are made smaller and more compact. Many engineers feel that this simplification has been achieved at the expense of one major flaw – the serial port is not available in this package. Otherwise, it is generally held to be *a good thing*.

The 40-pin DIL version shown in Figure 2.2a is the simplest version – i.e. an 8051/8751 (EP)ROM version with all four (8-bit) ports being used as I/O lines. Active-Low lines are identified with a preceding oblique, e.g. /EA.

('Active-Low' means that a logic 0 is needed for the input to perform its stated function. In this case /EA is pronounced 'not-External-Access' or 'not-E-A'. The implication is that in order to access external memory, this line should be tied low to 0 V.)

Common to all electronic circuits are the power supply lines V_{CC} and V_{SS}. In some variants, these may be called V_{DD} (= V_{CC}) or GND (= V_{SS}). Technically, this is a mix of terminology since V_{CC} is usually used for +V bipolar transistor supplies and V_{SS} is usually used for MOS Ground supplies. However, it is the terminology used by Intel. The letter 'C' in the designation implies a CMOS (low power variant). The power supplies needed are:

Voltage	Current
V_{CC} = +5 V_{SS} = 0 V	160 mA/20 mA (8031/80C31) or 250 mA/40 mA (8751/87C51)

The following section will cover the housekeeping input lines RST, XTAL1, XTAL2 and Vpp|/EA. (/PROG|ALE and /PSEN are outputs and will be explained later.)

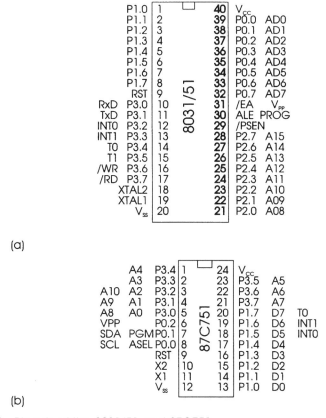

(a)

(b)

Figure 2.2 Pinouts of the 8031/51 and 87C751

Reset

RST is the system ReSeT line. It is essential to reset the microcontroller at power-up to initialize internal control registers. RST is an active high control line which will reset the microcontroller when a logic '1' is applied to it. The circuit of Figure 2.3 shows the basic configuration, with a switch added as an option to provide manual (additional) reset if required. The R-C circuit provides a time delay which must hold the RST line HIGH for a minimum of 24 oscillator periods.

The Watchdog Timer

In some designs it is considered necessary to add a WATCHDOG controller. This is a separate integrated circuit which will reset the microprocessor automatically if it is not regularly 'serviced' (send a logic signal from the microcontroller to reset the

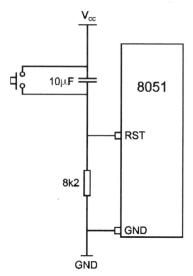

Figure 2.3 The ReSeT line

watchdog). It also provides the power-on-reset function of Figure 2.3. It is difficult to guarantee the integrity of any software, so if an error occurs, such as the microcontroller hanging up while waiting for an input which never arrives or getting stuck in a software loop, then the watchdog would not be serviced, so it would generate a reset.

Figure 2.4 shows one type of watchdog controller that integrates the reset function. Some members of the 8051 family have an inbuilt watchdog that can be programmed into the users' designs if necessary. These include the '542, '550, '552, '558, '562, '580 and '592 variants.

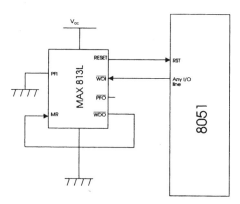

Figure 2.4 The watchdog timer

Clock and Resonator Circuits

All microprocessors require a clock. It is used to synchronize all the activities and data transfers within the CPU. Originally, the 8051 used quartz crystals up to 12 MHz, but later designs are capable of operating at up to 40 MHz. Figure 2.5(a) shows a circuit suitable for a 12 MHz crystal. This sounds quite a fast rate (40 MHz has period of some 25 ns) but the Intel family of devices divide this rate by 12 internally to produce the machine cycle – the basic interval which handles single-byte instructions. This brings the machine cycle rate down to 1 MHz (or up to 3.33 MHz for a 40 MHz clock).

If low cost or minimum component count is of prime importance, then consider the use of 3-terminal ceramic resonators which are made from barium titanate. These small devices have a central pin which is connected to ground. No capacitors are required. The disadvantages are:

- not as accurate as crystals (0.5% as compared to 0.002%)

- not as stable as crystals (as compared to 0.005%)

- not available in as high a frequency range as crystals (only up to 8 MHz).

Figure 2.5(b) shows a typical connection.

V_{PP}/EA is a control pin which has two functions. V_{PP} is a programming voltage for the EPROM version. A high voltage (12.5 V or 21 V, dependent on version) is used to store the users' programs on the internal EPROM. An 8751 programmer will apply address and data signals to the device, and a controlled pulse applied to the ALE pin will use the high voltage on the V_{PP} pin to store the data.

/EA is a signal which informs the MCS51 to look for its program in External Memory. If /EA is high (+5 V), the device looks for its program in internal EPROM.

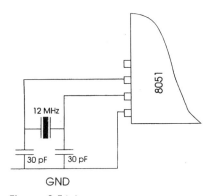

Figure 2.5(a)
A crystal resonator used as a 12 MHz clock

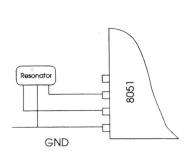

Figure 2.5(b)

If /EA is low (0 V) the device looks for its program in external EPROM. Several examples of applications of /EA will be given in the memory decoding section.

I/O Port Structures

All four ports in the MCS51 family are bidirectional. Each consists of a latch, an output driver, and an input buffer. When using the 8051/8751, all four ports can be used as inputs or outputs. Each I/O line can be independently used as an input or an output (although as an output, the pins can sink more than they can source). A source output is when current is output from the microprocessor (the microprocessor 'sources' current). A sink output is when current enters the output port (the microprocessor acts as a current 'sink'). Figure 2.6 shows the difference.

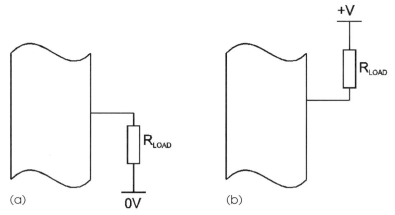

Figure 2.6   (a) Output source; (b) output sink

Ports 1, 2 and 3 are called quasi-bidirectional. After the 8051 has been powered up or after a RESET, all ports act as inputs. The exception to this is when the EA (External Access) pin is connected to 0 V in order to instruct the IC to access external program memory. In this case, Ports 0 and 2 have special functions which I will deal with shortly. The quasi-bidirectional structures are shown in Figure 2.7. Ports 1 and 3 (and Port 2 if not used to access memory) have a 'weak internal pull up' formed from a FET switch. When used as an output, each pin of each of these ports will turn on the lower FET to output a '0' or it will turn off the lower FET to output a '1' via the pull-up resistor. This is the reason why the 8051 ports can sink more current than they can source. The lower FET represents a low impedance connection to ground when it is turned on, while the pull-up resistor (FET) is a higher impedance. The diagrams are actually a little oversimplified. In parallel with the pull-up 'resistor', there is another FET which switches on for 2 oscillator periods whenever a '1' is to be output. This FET pulls the output voltage up sharply, i.e. if

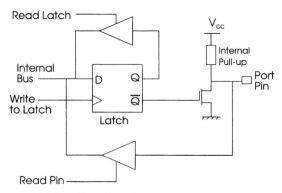

Figure 2.7 A quasi-bidirectional port

there is a capacitive load, it would be rapidly charged to the V_{CC} potential and would stay there once the weak pull-up was the only output device.

This facility is useful, because if you wish to create an input from (say) a push-to-make push-button, then the only circuit needed is that of Figure 2.8. Reading this pin would normally result in a '1' being input since the pull-up resistor would connect that voltage to the CPU buffer. Pressing the button would connect the pin to ground and hence a 0 V would be input.

When using the pin as an output, the only care that must be taken is that you must not write a '0' out on that pin before trying to use it again as an input. To do so would cause a conflict with any devices which may be connected to the pin. If an external device is trying to input a '1' and the 8051 is simultaneously trying to output a '0', then catastrophe will be the only result! The output FET of the 8051 is capable of sinking up to 0.5 mA which could damage sensitive devices which are trying to output a logic '1'. Alternatively, if the external device trying to put a signal into the 8051 is capable of outputting more than 15 mA, then there is a good chance of damaging the input buffer of the 8051. In practice, though, it is rare to want to use a port as both input and output, so this situation should not arise.

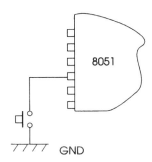

Figure 2.8 Single push-button input

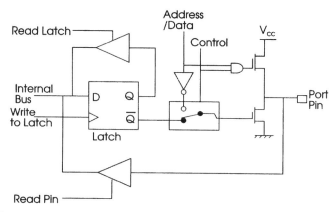

Figure 2.9 Circuitry of Port 0

Port 0 is different. This port has no weak pull-up because of its capability to read from and write to external memory devices. It has two output FETs which provide high current paths to either the $+V_{CC}$ or GND supplies. The pull-up FET only operates when executing an instruction which causes it to access external memory. The output pins for normal operation are 'open-drain', i.e. only connected to the lower pull-down FET. It would need external pull-up resistors to function in the same way as Figure 2.7. It is a true bidirectional port because when used as an input, the pins float (are not connected to any particular voltage).

Some Design Examples

Figure 2.10 is a simple example of an 8751 being used as a combinational door lock controller. Port 1 is used to operate a 16-key matrix keypad, while Port 2 operates the indicator lights and door solenoid.

The keypad is arranged in columns and rows. Pressing a key causes a column to be connected to a row. The software of the MCS51 must sequentially 'strobe' the four columns. This means that it outputs a logic '1' pulse to column 1; then a logic '1' pulse to column 2 and so on. While a column is driven high by a logic '1', the microprocessor inputs the data on the 4 rows. If a button in that column is pressed, it will be sensed in the appropriate row.

In Chapter 4 (page 113) a dedicated keyboard interface IC is introduced. This is the 74C922. The design in Figure 2.10 uses 8 pins connected to the microcontroller, while the 74C922 would only connect 5 pins to the IC. Chapter 4 (page 113) explains how to use an output driver IC to increase the current available from the microprocessor.

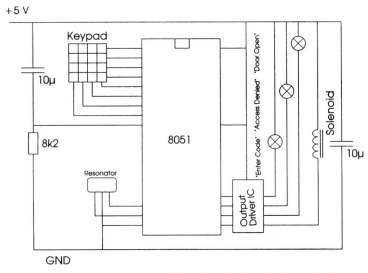

Figure 2.10 An 8751 used as a combinational door lock controller

Accessing External Memory

If the /EA line of the microcontroller is LOW, then Port 0 and Port 2 are used to access the program memory. Every time a byte of data is to be fetched from this external memory the following sequence occurs:

1. Port 0 outputs the low byte of the external memory address while Port 2 outputs the high byte of the external memory address.

2. The ALE line (pin 30) is asserted – this will cause the lower byte of the address to be stored by an external latch (see Figure 2.11).

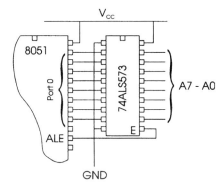

Figure 2.11 The ALE line (pin 30) is asserted, causing the lower byte of the address to be stored

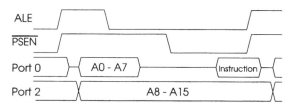

Figure 2.12   Timing of the control signals

3. Port 0 is 'floated' (put in a high impedance mode).

4. /PSEN (pin 29) is asserted. This signal operates the external memory device (i.e. /CE of the EPROM).

5. The byte of data is then read in by Port 0.

This is somewhat confusing as a word description, so Figure 2.12 shows the timing of the control signals. The sequence of these signals (ALE and /PSEN) is started automatically by the action of setting /EA low. Additional circuitry would be needed to use these signals and Figure 2.14 shows how this could be achieved. The latch could be a '373 or '573 octal latch. These devices store a byte of data when the E (enable) line is activated.

Logical Separation of Program Memory (read only) and Data Memory (read/write)

An original design feature of the 8051 is that it has separate Program and Data Memory. The microcontroller uses /PSEN (pin 29) to access Program Memory and /RD (pin 17) and /WR (pin 16) to access Data Memory. The logical separation of these two memory spaces allows the Data Memory to be accessed by 8-bit addresses which can be more quickly stored and manipulated by an 8-bit CPU (although 16-bit addresses can be used if necessary).

Figure 2.13 shows the timing of data transfers, while Figure 2.14 shows how the 8031 could support both Program Memory in EPROM and Data Memory in RAM. The distinction is due to the fact that the /PSEN signal is used to access the EPROM, while the /RD – /WR signals are used to access the RAM.

Larger size EPROM and RAM devices can be used directly up to 64 kbytes each. Designers of embedded systems are frequently as interested in target board size as much as cost, so the usual solution is to use as large a memory device as possible, i.e. use 1×64 kbyte instead of 4×16 kbyte. However, Figure 2.15 shows a fully decoded example with segmented memory and memory-mapped I/O.

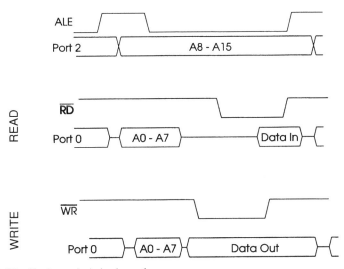

Figure 2.13 Timing of data transfers

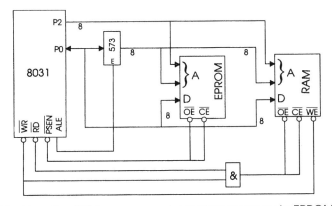

Figure 2.14 How the 8031 can support program memory in EPROM and data memory in RAM

A single 16 kbyte memory will have 14 address lines which address

```
from  00  0000  0000  0000  binary  0 000  hexadecimal
to    11  1111  1111  1111  binary  3 FFF  hexadecimal
```

There are four of these shown in Figure 2.15 and there are a total of 16 address lines of which only 14 are used on each RAM IC. The upper two address lines allow the 74138 to work out which of the RAM IC's are to be accessed for any particular address. The 74138 is a 3 to 8 line decoder. In this circuit, only two binary inputs are used to determine which of the four chip enable (CE) outputs activate which RAM device.

A_{15}	A_{14}	R_4	R_3	R_2	R_1
0	0	1	1	1	0
0	1	1	1	0	1
1	0	1	0	1	1
1	1	0	1	1	1

In this particular example of Figure 2.15, the 8255 I/O port responds to one of four RAM addresses. That is, it appears as if it is 4 of the 64 kbyte memory addresses. To do this, the particular 4 addresses must be detected by the address decoder, and when it enables the 8255 I/O port, it must disable those particular 4 RAM addresses in the RAM devices. The 8255 I/O port has 3×8 bit ports A, B and C; so it expands the I/O capability of the microprocessor. The reason it needs 4 discrete addresses is that it has 3 I/O ports and 1 control port. This is an example of 'memory mapped I/O' in that the I/O device behaves somewhat like a memory device in that it can be written to and read from.

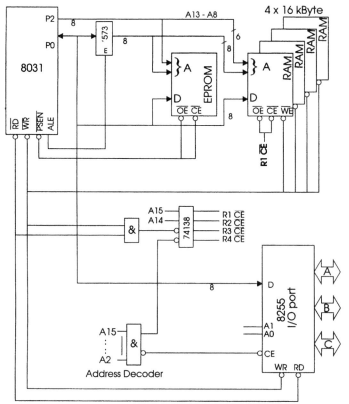

Figure 2.15 An 8031 with segmented memory and memory-mapped I/O

Non-volatile Memory

There are many situations where it is important not to lose the contents of memory when the power is removed. The two main solutions for this use Electrically Erasable EEPROM or battery/capacitor-backed RAM.

EEPROM is available in serial or parallel access. Serial EEPROM needs special algorithms to save and read data. Parallel EEPROMs are easier to drive since they more closely resemble conventional EPROMs.

RAM whose contents are preserved with backup capacitor or battery are most conveniently interfaced via proprietary control ICs. The difficulty with RAM is that if there is any noise or jitter on the /WE and /CE lines as the main system power shuts down, then parts of memory will be lost. Several manufacturers produce suitable ICs which can be used, as in Figure 2.16. This circuit is also capable of driving the RESET line of the 8031.

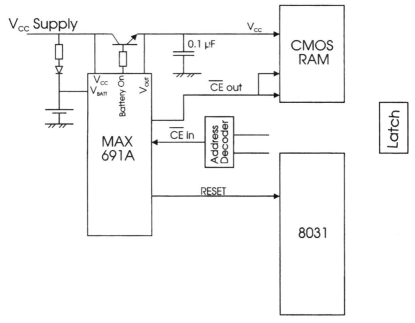

Figure 2.16 Control circuitry for battery-backed RAM

Some applications such as data loggers could require much more memory than the 64k directly available from the 8051. Figure 2.17 shows a circuit which uses spare lines of Port 3 to activate different banks of RAM. The 3 to 8 line decoder 74LS138 allows up to 8×64 kbytes of memory to be selected. This arrangement could be extended almost indefinitely. In practice, of course, PLDs (Programmable Logic Devices) would be used since they could integrate many other functions.

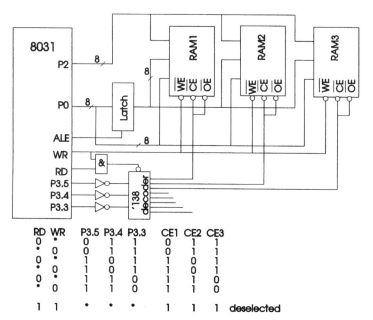

Figure 2.17 Using spare lines of Port 3 to activate different banks of RAM

Port 3 and two lines of Port 1 have alternate functions. These can be used selectively (see Table 2.2).

Intel use the dot notation of the form P3.4 to indicate bit 4 of Port 3.

Table 2.2

PORT PIN	Alternate Function
*P1.0	T2 (Timer/Counter 2 external input)
*P1.1	T2EX (Timer/Counter 2 Capture/Reload trigger)
P3.0	RXD (serial input port)
P3.1	TXD (serial output port)
P3.2	/INT0 (external interrupt 0)
P3.3	/INT1 (external interrupt 1)
P3.4	T0 (Timer/Counter 0 external input)
P3.5	T1 (Timer/Counter 1 external input)
P3.6	/WR (external data write strobe when /EA is low)
P3.7	/RD (external data read strobe when /EA is low)
*P1.0 and P1.1 serve these alternate functions only on the 8052 and more complex MCS family parts which have more than just the two timers.	

Questions

1. Design the hardware of an 80C51 based electronic die where the input is a single 'throw' switch and the output is the drive to a seven-LED array. How could this be reduced to 4 outputs?

2. Show how to interface an 8031 to a 27128 EPROM, a 62128 Static RAM and 2×8255 parallel input output interface devices. The 8255 devices should use addresses $FF00_H$–$FF03_H$ and $FF04_H$–$FF07_H$.

3. Look forward to page 126; the ADC 0803 analogue to digital converter. Show how to design this into the 8031 based system of question 2 so that the address of the ADC is $F000_H$.

Architecture

We cannot make much more forward progress with the device until more is revealed about the internal architecture of the 8051. Figure 2.1 shows a typical implementation with the four ports, two timers, internal registers and all the other bits and pieces which make up the microcontroller. The base model has 128 bytes of available RAM addressed from 00 to 7F. There is an additional area of RAM from 80 to FF called the Special Function Register (SFR) space which hold registers which control or report on the many functions built into the IC. These two memory areas are called the direct memory space. All of these addresses can be read or written to, but obviously the SFR space is reserved for the special functions. Not all of the 127 locations in SFR are used.

Register Structures

There are 8 general purpose registers which are called R0, R1, R2, R3, R4, R5, R6 and R7. They are not physical registers in the CPU as such. They actually occupy general-purpose RAM addresses 00–07. The 8051 allows the programmer to have 4 sets of these 8 registers. These are called Register Banks 0, 1, 2 and 3, but at any one time there are only 8 active registers called R0–R7. At power-up, the machine defaults to Register Bank 0.

Table 2.3

Register Bank	Address in RAM of R0–R7
0	00–07
1	08–0F
2	10–17
3	18–1F

Table 2.4

Direct Byte Address	Hardware Register Symbol	Function
F8H		
F0H	B	B Register
E8H		
E0H	A or ACC	A Register
D8H		
D0H	PSW	Program Status Word
C8H		
C0H		
B8H	IP	Interrupt Priority
B0H	P3	Port 3
A8H	IE	Interrupt Enable
A0H	P2	Port 2
99H	SBUF	Serial Data Buffer
98H	SCON	Serial Control
90H	P1	Port 1
8DH	TH1	Timer 1 High Byte
8CH	TH0	Timer 0 High Byte
8BH	TL1	Timer 1 Low Byte
8AH	TL0	Timer 0 Low Byte
89H	TMOD	Timer Mode
88H	TCON	Timer Control
83H	DPH	Data Pointer High Byte
82H	DPL	Data Pointer Low Byte
81H	SP	Stack Pointer
80H	P0	Port 0

So, if Register Bank 0 is selected, storing a number in Register R3 would cause it to be saved at internal RAM location 03. But if Register Bank 1 is selected, storing a number in Register R3 would cause it to be saved at internal RAM location 0B. Obviously, there is a potential problem here which could be caused by moving data from a Register when the wrong Register Bank is selected. The wrong data would be moved. This is the responsibility of the programmer. (The addresses and data shown are in the hexadecimal (16) number base. For an explanation of this, see Appendix B.)

As far as the Special Function Register (SFR) is concerned, Table 2.4 shows the various functions of the Direct Bytes.

The **Hardware Register Symbol** is the term for each register coined by Intel and it is also the term which is accepted by all assembler software (which converts the 8051 instructions into machine code). The address is just as acceptable a term as the symbol, so instructing the CPU to write (move) the hexadecimal number 3A to address 80 would be achieved with the instruction

```
mov 80H, #3AH
```

More readable would be the instruction

```
mov P0,  #3AH
```

In both cases, the binary number 00111010 would be written out on Port 0. Most assemblers use the suffix H to indicate hexadecimal numbers, B for binary numbers (and sometimes O for octal numbers). However, there are a few which use an H (or B or O) prefix with the number itself in single quotes, i.e. H'3A'. Sometimes when handling control registers, it helps fix the bit pattern more visually if binary is used: as in

```
mov P0,  #00111010B
```

The # symbol is used by Intel to indicate that it is the number itself which is to be loaded. The difference would be

```
mov P0,  #00111010B   ; output 00111010 to Port 0
mov      P0, 00111010B; output the contents of
                      ; location 00111010 (3A) to
                      ; Port 0
```

The full instruction set is shown in Appendix A. Note that any register in the SFR which has an address ending in -8 or -0 is bit addressable. This means that the dot notation can be used to read or write to an individual bit in that register. The following are valid instructions:

```
setb    P0.4        ; Setting bit 4 of output port P0
clr     IE.0        ; Clearing bit 0 of the
                    ; Interrupt Enable Register
jb      a.0, next   ; jump if bit 0 of the Accumulator =
                    ; 1 to the line in the program
                    ; which is identified with
                    ; the label 'next'
```

If the register is not bit addressable, individual bits can be manipulated with the logical OR or AND instructions, e.g.

```
orl TMOD, #01000000B   ; set bit 6
                       ; orl = logical OR
anl TMOD, #10111111B   ; reset bit 6
                       ; anl = logical AND
                       ; take care where and how you
                       ; pronounce them
```

Appendix B explains more about hexadecimal numbers and arithmetical/logical manipulation of this sort.

The other registers are all special function and are:

A register or accumulator. This register has its counterpart in most 8-bit microprocessors. It is connected to the output of the ALU (Arithmetic and Logic Unit) and is the repository for the result of most arithmetic and logic operations. It is also the source of data for many instructions.

Table 2.5

PSW.7	PSW.6	PSW.5	PSW.4	PSW.3	PSW.2	PSW.1	PSW.0
CY	AC	F0	RS1	RS0	OV	–	P

CY	PSW.7	Carry Flag. Set/Cleared by hardware or software during certain arithmetic and logical instructions
AC	PSW.6	Auxiliary Carry Flag. Set/Cleared by hardware during addition or subtraction instructions to indicate carry or borrow out of bit 3
F0	PSW.5	Flag0. Set/Cleared/Tested by software as a user-defined status flag
RS1	PSW.4	Register Bank select bit 1
RS0	PSW.3	Register Bank select bit 0
OV	PSW.2	Overflow Flag. Set/Cleared by hardware during arithmetic instructions to indicate overflow conditions
–	PSW.1	(reserved)
P	PSW.0	Parity Flag. Set/Cleared by hardware each instruction cycle to indicate an odd/even number of 'one' bits in the accumulator

Table 2.6

RS1 PSW.4	RS0 PSW.3	Register Bank Selected	Internal RAM Address
0	0	0	00–07
0	1	1	08–0F
1	0	2	10–17
1	1	3	18–1F

B register. This is mostly used in multiply and divide operations. If not required for these functions, the memory space is available as a general-purpose 8-bit register.

PSW (Program Status Word). The Status register exists in one form or other for all microprocessors. It reports on both the nature of the result of some arithmetic and logic operations and the state of the A register. It allows the programmer to select which of the register banks is in use.(See Table 2.5.)

The working Register Bank would be selected according to Table 2.6.

One way to select a register bank would be to write directly to the PSW:

```
mov psw, #00010000B  ; Select Register Bank 2
```

This is OKish, but it does override some of the status bits. A better way would be to alter the appropriate PSW bits directly as in:

```
setb psw.4           ; Set bit 4
clr  psw.3           ; Clear bit 3
```

or even by:

```
anl  psw, #11100111B ; Select Register Bank 0
                     ; (clear bits 4 and 3 first)
orl  psw, #00010000B ; Select Register Bank 2
```

Most assemblers will even recognize the abbreviations RS1 and RS0 allowing

```
setb rs1
clr  rs0
```

which is even more meaningful.

Some instructions test bits in the PSW:

```
jc next                   ; test the overflow bit (carry)
                          ; and jump if set to label 'next'
```

Table 2.7

IP	Interrupt Priority	This sets the priority of the 5 interrupt sources
P3	Port 3	A general-purpose Input/Output port. It also has alternate functions which allows it to act as I/O for Serial, Timer and External Interrupt signals
IE	Interrupt Enable Register	
P2	Port 2	Also used to provide the upper byte of external memory address
SCON	Serial Port Control/Status Register	
P1	Port 1	
TCON	Timer/Counter Control/Status Register	
SP	Stack Pointer	
P0	Port 0	

The following (trivial) program shows how the input from Port 3 could be copied onto Port 1.

```
        org 0

loop    mov A,    P3
        mov P1,   A
        jmp loop

        end
```

This shows the assembler directives org and end.

Org   tells the assembler to assemble code from memory location 0000. This is where the CPU looks for its first instruction after a RESET

end   instructs the assembler to stop assembling.

In this example, it looks as if a simpler solution would be if there was such an instruction as mov P1, P3. This does not exist. Some microprocessor families do allow such a flexibility, but the 8051 keeps a manageable instruction set which is more easily learnt, even if occasionally, it is at the expense of an extra line or two of code.

Questions

1. Write a program to set the carry flag if the contents of register R1 is between the values held in registers R2 and R0 (contents of R2 > contents of R1 > contents of R0).

2. Write a program to convert a 16 bit number stored in registers R0 and R1 into 4 ASCII digits stored in registers R3, R2, R1, R0. Page 189 shows the ASCII values of numbers.

3. Write a program which converts a binary number in the range 00–99 (00000000–01100011) into binary coded decimal, i.e. as explained on page 187 so that the expected range is 00000000–10011001. Assume that the number is initially in the R1 register and that the answer must be left in the R2 register.

Timer/Counters

The 8051 has two 16-bit Timer/Counter registers – Timer 0 and Timer 1. When activated, these registers increment once for each event. An event can be related to the system clock (Timer) or from an external logic signal transition (Counter). If required, a Timer/Counter register can be set up to generate an interrupt when it 'rolls over' from all 1s to all 0s, i.e. when it reaches its full possible count of FFFFH (65535) and advances the register to 0000 on its next increment. This Timer/Counter register is called TL0 and TH0 for Timer 0 or TL1 and TH1 for Timer 1. The 'L' and 'H' refer to the low and high byte of the 16-bit register respectively.

The Timer/Counter register can be preset to any value so that any time or count limit can be set up. For example, to time or count 20 events, the register should be loaded with the number FFDCH (65616). There are several modes of operation of the Timer/Counters. The hardware inputs for the Timer/Counter channels 0 and 1 are Port pins P3.4 and P3.5 respectively.

When used as a timer, deriving its count from the system clock, it first divides that clock by 12 as in Figure 2.18.

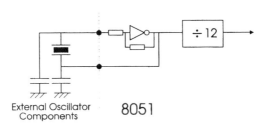

External Oscillator Components   8051

Figure 2.18   Dividing the system clock by 12

Table 2.8

Timer Mode Control Register			
TIMER 1	TMOD.7	GATE	When gate = 1, Timer 1 is enabled only when INT1 = 1 and TR1 = 1
	TMOD.6	C/T	Counter/Timer select (0 = Timer, 1 = Counter)
	TMOD.5	M1	mode
	TMOD.4	M0	mode
TIMER 0	TMOD.3	GATE	When gate = 1, Timer 0 is enabled only when INT0 = 1 and TR0 = 1
	TMOD.2	C/T	Counter/Timer select (0 = Timer, 1 = Counter)
	TMOD.1	M1	mode
	TMOD.0	M0	mode

Table 2.9

M1	M0		Operating Mode
0	0	mode 0	13-bit Timer/Counter
0	1	mode 1	16-bit Timer/Counter
1	0	mode 2	8 bit auto reload Timer/Counter
1	1	mode 3	Timer/Counter stopped

The operation of Timer 0 and Timer 1 is effected by

Timer Mode Control Register	TMOD	(address 89)
Timer Control Register	TCON	(address 88)

Mode 0

This is a (virtually obsolete) mode which is included for functional compatibility with the earlier 8048 microprocessor family. In it, both Timer 0 and Timer 1 are 13-bit Counter/Timers with a divide-by-32 prescaler. The 13 bits of the Timer/Counter register are specifically:

- the 8 bits of TH0 (or TH1)

- the lower 5 bits of TL0 (or TL1).

Table 2.10

Timer Control Register		
TCON.7	TF1	Timer 1 overflow flag. Set when Timer/Counter overflows. Cleared by hardware when the processor vectors to the interrupt routine
TCON.6	TR1	Timer 1 run control. Set or cleared by software to turn the Timer/Counter on or off
TCON.5	TF0	Timer 0 overflow flag. Set when Timer/Counter overflows. Cleared by hardware when the processor vectors to the interrupt routine
TCON.4	TR0	Timer 0 run control. Set or cleared by software to turn the Timer/Counter on or off
TCON.3	IE1	Interrupt 1 edge flag. Set by hardware whenever an external interrupt is detected. Cleared when the interrupt is processed
TCON.2	IT1	Interrupt 1 Type control bit. 0 = low level input trigger 1 = falling edge input trigger
TCON.1	IE0	Interrupt 0 edge flag. Set by hardware whenever an external interrupt is detected. Cleared when the interrupt is processed
TCON.0	IT0	Interrupt 0 Type control bit. 0 = low level input trigger 1 = falling edge input trigger

The counted input is enabled to the Timer when

- timer 0:   TR0 = 1   **AND**   (GATE = 0 OR INT 0 = 1)

- timer 1:   TR1 = 1   **AND**   (GATE = 0 OR INT 1 = 1)

If GATE = 1 then the Timer can be controlled solely by the external input INT to facilitate pulse width measurement.

```
mov TMOD, #10000000B   ; both channels set into mode
                       ; 0 . . . both are timers.
                       ; Timer 1 is started by pin 13
                       ; (port3.3) going high
setb TR0               ; Timer 0 is started by TR0
                       ; control being set.
setb TCON.4            ; Alternative start to Timer 0
```

Mode 1

This is the same as Mode 0, except that the Timer register is run with all 16 bits of TL0 and TH0 (or TL1 and TH1).

```
mov  TMOD,  #10010001B
```

In this example, both channels are set up to be Timers in Mode 1. Timer 1 is to be controlled by the external pin 13, P3.3 while Timer 0 is to be controlled by the software setting of TR0. In Mode 1, both of the timers are 16 bit and so TH and TL must be loaded. Timer 1 is to be loaded with F000 and Timer 0 with D000.

```
mov  TH1,  #F0H  ; Timer 1 register initialisation
mov  TL1,  #00H
mov  TH0,  #D0H  ; Timer 0 register initialisation
mov  TL0,  #00H
```

The TCON timer control needs setting up, but in this example it is not proposed to use interrupts, so it becomes a simple task:

```
mov  TCON,  #00000000B
```

Thereafter, Timer 1 can be started with a signal on External Interrupt 1 while Timer 0 can be started with an instruction such as

```
setb TR0
```

For polling purposes, each byte of the Timer/Counter registers would have to be monitored. The following code locks up progress until both High and Low bytes of Timer 0 advance to zero.

```
wait1:   cjne TH0, #00, wait1   ; if TH0 is not =
                                 ; 0 jump to wait1
         cjne TL0, #00, wait1   ; if TL0 is not =
                                 ; 0 jump to wait1
```

The first line compares the contents of Timer register TH0 with 00 and jumps if it is *n*ot *e*qual (to 00) to the line with the label wait1. The second line compares the contents of the Timer register TL0 with 00 and jumps if it is not equal (to 00) to the line with the label wait1. The overall effect is that the code will 'hang up' on these two lines until the 16-bit register TH0 and TL0 advances from its initial setting of D000 to 0000. Polling is cumbersome and wasteful of time, so all but the most trivial programs tend to use interrupts. It might be risky for the externally controlled Timer 1 to hang up in such an endless loop in case the external signal was lost or changed unexpectedly.

If the MCS51 operates from a 12 MHz clock, then timer 0 would increment every 1 ms once setb TR0 has been issued.

Mode 2

This useful mode configures the Timer register as an 8-bit Counter with automatic reload. When TL0 (or TL1) overflows, it sets the Timer Flag TF0 (TF1) and reloads TL0 (TL1) with the contents of TH0 (TH1). The reload leaves TH0 (TH1) unchanged. This mode is quite often used for baud rate generation for the serial communication port. Or, for any application which requires a regular frequency to be output. When setting up, load both TH0 and TL0 (TH1 and TL1) with the reload figure since the TL register is only reloaded when it overflows. Otherwise, the first timing run could be slightly longer if the TL register is at its reset value of 00.

Mode 3

This mode affects the two timers in different ways.

Timer 1 in Mode 3 simply stops counting. The effect is the same as setting TR1 = 0.

Timer 0 in Mode 3 splits the TL0 and TH0 registers into two independent 8-bit counter timers. TL0 is controlled in exactly the same way as in Mode 1, i.e. using control bits C/T GATE and TR0. TH0 can only act as a timer and uses the Timer 1 control functions TR1 and TF1. Timer 1 loses its ability to generate interrupts if Timer 0 is set into Mode 3. However, this mode finds a use in the circumstances when an extra Timer function is required.

Timer Example: A 50 Hz Square Wave

The following example shows how to generate a 50 Hz square wave on Port 1 bit 0. 50 Hz requires a 10 ms on and 10 ms off-cycle which can be derived from a 12 MHz clock if 10 000 machine cycles are counted for each ON and each OFF period. To make a timer count up to 10 000, the 16-bit register must actually be loaded with 65 536 − 10 000 = 55 536 − which is D8F0$_H$.

```
; 50 Hz clock from a 12 MHz crystal

org   0

mov   TH0,   #D8H
mov   TL0,   #F0H              ; set up register counts
mov   TCON,  #0                ; no interrupts
```

```
mov   TMOD,  #00000001B        ; timer 0 in mode 1
setb TR0                       ; start timer 0

label:
cjne TL0, #00, label           ; wait until low byte
                               ; overflows
cjne TH0, #00, label           ; wait until high byte
                               ; overflows

mov   TH0,    #D8H             ; reset counter high
                               ; byte
mov   TL0,    #F0H             ; reset counter low byte
cpl   P1.0                     ; change state of the
                               ; output pin

jmp label                      ; loop to do it all
                               ; again

end
```

Questions

1. Two tone horn. Write a program which outputs two tones from P3.1 when the single P3.0 input goes high. The two tones should switch between 256 Hz and 440 Hz at a rate of 0.6 second intervals.

2. An alarm annunciator has 3 inputs on P1.0, P1.1 and P1.2. These are labelled and function as ALARM, ACCEPT, CLEAR. There are two outputs on P1.6 and P1.7 which provide a drive to a horn and a lamp. When the ALARM input goes high, the horn should sound a 1 kHz tone with an off–on pattern of 2 Hz. The output lamp should flash at a sound synchronized rate of 2 Hz. When ACCEPT is pressed, the alarm is cancelled and the lamp stays steady-on. When CLEAR is pressed, the system is reset with both outputs being set to '0'.

3. An 8051 has two inputs on P2.0 and P2.1, and uses Port 3 as an output and P2.7 as an 'over-range' output. The microcontroller is to time the difference between input P2.0 going high and input P2.1 going high. The basic time interval is to be LSB = 25 ms (LSB = least significant bit). The output should be displayed in binary from Port P3. If the delay exceeds the maximum counts, the over-range output should go high for 2 seconds before the system resets.

4. Write a program which outputs the frequency of a stream of pulses. Assume that the signals are input on Port P1.0. Start a timer when the input first performs a 0–1 transition, and stop the timer when the input next does a 0–1 transition. Then use the mathematical DIVIDE function to calculate the frequency of the input. For simplicity, assume that the timer never counts more than 255 between pulses

(an 8 bit timer). How could this problem be extended to cope with the situation where the input stream counts up to 65535 between input pulses (a 16 bit timer?). More generally, how could the problem be extended to cope with any time interval between pulses. How could you set a 'time-out' to terminate the waiting after a sensible interval?

Interrupts

A CPU executes a program instruction by instruction. It will work its way through the instructions sequentially unless one of two events occurs.

1. The current instruction is an explicit command to jump a set number of instructions or to jump to a particular address.

2. An interrupt occurs. If interrupts have been enabled then that interrupt will cause the CPU to jump to a fixed location.

The 8051 has a useful vectored interrupt system that allows for 5 interrupt sources which are (in the order of priority) shown in Table 2.11.

The 'vector' address represents the address to which the CPU will relocate in the event of the named interrupt occurring. What will actually happen is that the Program Counter register holds the address of the next instruction to be executed. When the interrupt occurs, the Program Counter saves that next address on the memory area known as the *stack* and loads one of the above addresses into the Program Counter instead. If, for example, IE1 (External Interrupt 1) had been enabled and at a particular point in the program, Port 3 bit 3 (pin 13) was taken from a logic 1 to a logic 0, then an interrupt would be generated. The Program Counter would be loaded with the address 0013 and after the current instruction had finished, then the CPU would fetch its next instruction from location 0013. Each interrupt has been allocated 8 bytes, and so if the interrupt is to take more code than this, then the simplest action is to place a jmp (jump) instruction at address 0013.

Table 2.11

Interrupt	Flag Name	Vector
External Interrupt 0	IE0	0003
Timer/Counter 0	TF0	000B
External Interrupt 1	IE1	0013
Timer/Counter 1	TF1	001B
Serial Port	RI + TI	0023

The code which handles the interrupt is commonly called the interrupt service routine. The end of the interrupt service routine is terminated with the instruction reti (**RET**urn from **I**nterrupt). This retrieves the return address from the stack and loads it into the Program Counter. Hence, the CPU resumes with the next instruction it was due to execute before the instruction occurred.

One last point. The vectors for the interrupt service routines are located fairly low in memory. If an interrupt doesn't happen, then after a power-up or system reset, the CPU will quite naturally execute code from locations 0000, 0001, 0002, 0003, 0004, 0005, etc. It will quite happily work through the addresses, not knowing that the code at location 0003 contained code for External Interrupt 0. The way around this is to start the program something along the lines of

```
org             0000
ajmp            start           ; bypass vectors
org             0003H
ajmp            isr_ie0         ; external interrupt 0
org             000BH
ajmp            isr_tf0         ; timer 0 interrupt
org             0013H
ajmp            isr_ie1         ; external interrupt 1
org             001BH
ajmp            isr_tf1         ; timer 1 interrupt
org             0023H
ajmp            isr_ti_ri       ; serial interrupt

start:                          ; first piece of code
```

Where, for example, jmp isr_tf1 represents an instruction to jump to the piece of code labelled isr_tf1.

Here is a simple program which is designed for an 8051 operating from a 12 MHz clock and merely outputs a slow square wave from Port 1.0.

```
start:      .equ  0000h
timer0:     .equ  000Bh
.org        start
jmp         main

.org        timer0              ; timer vector
jmp         timer_routine

main:
mov         sp, #20h            ; move stack to allow
                                ; use of register bank 1
setb        rs0                 ; register bank 1
mov         r7, #0              ; use r6 & r7 to count
                                ; the number of
                                ; interrupts
```

```
        mov       r6, #0
        clr       rs0                    ; register bank 0

                  ;*** set up timer registers ***
        mov       tmod, #00000010b       ; auto reload TH to TL
        mov       th0, #6                ; 250 counts @ 12 MHz =
                                         ; 250 µS
        mov       tl0, #6
        setb      et0                    ; enable timer
                                         ; interrupts
        setb      ea                     ; enable interrupts
        setb      tr0                    ; start count
        jmp       $                      ; hang up until
                                         ; interrupts occur
                                         ; jmp $ means jump to
                                         ; the current line
                  ; *** interrupt service routine ***
timer_routine:
        setb      rs0                    ; select second register
                                         ; bank
        inc       r7
        cjne      r7, #40, timer_end     ; count 40 register
                                         ; R7 increments
        mov       r7, #0
        inc       r6
        cjne      r6, #100, timer_end    ; count 100 register
                                         ; R6 increments
        mov       r6, #0
        cpl       p1.0                   ; or whatever code is
                                         ; needed – this toggles
                                         ; the output state of
                                         ; output port in P1.0 at
                                         ; 1 Sec intervals
timer_end:
        clr       rs0
        reti

        .end
```

Some words must shortly be said about the nature of the 8051 stack, but first we must look at the interrupt control structure more closely. Special Function Register IP affects the priority of the interrupts. It was stated initially, that the priority of interrupts was IE0, TF0, IE1, TF1, TI + RI.

So, if all interrupts have been enabled, then if the CPU is handling <external interrupt 1> and a <timer 0> interrupt occurs, the CPU will:

- suspend execution of `isr_ie1` code

- deal with `isr_tf0` code and when it has finished

- return to where it was, in this case to `isr_ie1` code.

However, since it has a lower priority, if a <timer 1> event occurs while the CPU is handling <external interrupt 0> then the code for `isr_tf1` will just have to wait until the CPU has finished with `isr_ie0`. One way to alter the priority is to use the IP register. This has 5 active bits which alter the priorities according to:

IE0	IP.0	External Interrupt 0
TF0	IP.1	Timer 0 Interrupt
IE1	IP.2	External Interrupt 1
TF1	IP.3	Timer 1 Interrupt
RI + TI	IP.4	Serial Port Interrupt

An instruction such as `mov ip, #00001001B` would set TF0 and RI+TI as high priority interrupts. The priority order now becomes:

TF0
RI + TI
IE0
IE1
TF1

Figure 2.19 shows this diagrammatically:

To enable interrupts, the Interrupt Enable (IE) register must be used.

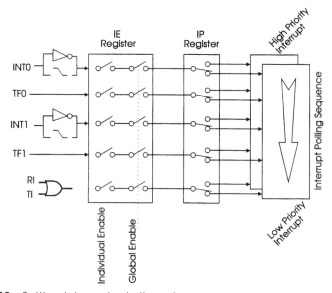

Figure 2.19 Setting interrupt priority order

Table 2.12

EA	IE.7	Enable All control bit	0 = all interrupts disabled
	IE.6		
	IE.5		
ES	IE.4	Enable Serial port	1 = enable serial interrupts
ET1	IE.3	Enable Timer 1	1 = enable Timer 1 interrupts
EX1	IE.2	Enable External 1	1 = enable external 1 interrupts
ET0	IE.1	Enable Timer 0	1 = enable Timer 0 interrupts
EX0	IE.0	Enable External 0	1 = enable external 0 interrupts

```
mov  ie, #10001010B      ; enable timer 0 and
                         ; timer 1 interrupts
clr  ea                  ; clear ie.7 - disable
                         ; all interrupts
```

Stack

At a reset, the stack pointer is set to 07. This is an 8-bit register which is used for holding the address of a position in memory for temporary storage of data and addresses. When data is pushed onto the stack, the stack is first incremented then data is stored at that stack pointer address. A pop retrieves data from the current stack pointer address and then decrements the stack pointer. So, the first interrupt to occur would store the return address for CPU code at locations 08 and 09. Note that the default setting for the stack will cause it to overwrite the register bank 1. (Recall that Register Bank 1 sets R0–R7 as using RAM memory addresses 08–0F.) It does not overwrite the interrupt vector addresses because of the way in which the 8051 separates program code and data. Interrupt vectors would be part of the program code, whilst register banks are RAM.

The usual thing to do in any non-trivial application is to adjust the stack location, e.g.

```
mov  sp, #30H  ; relocating the stack to start at
               ; RAM address 30
```

The stack grows as necessary. Every push instruction or subroutine call or interrupt causes the next address(es) in the stack memory to be used. The limit to the size of the stack is the available data RAM.

There is an interesting modification suggested by Philips Semiconductors which allows the 8051 to seem to have 5 external interrupts. EI0 and EI1 are already known about. Two others can be created by setting Timer/Counters 0 and 1 into 8-bit

Counter mode (Mode 2) with a preloaded count of FF. Thus, one more falling edge input on the Timer 0 or Timer 1 input would cause the Counter to roll over and generate an interrupt. The interrupting Counter would automatically be reloaded with FF from the TH0 or TH1 register so that the interrupt could occur again as soon as the interrupt service routing had been completed. To set up both timers for this type of operation,

```
mov    TMOD, #01100110B        ; both timers in mode 2
                               ; in counter mode
                               ; no gate control
setb  TR0                      ; start timer 0
setb  TR1                      ; start timer 1
setb  ET0                      ; enable timer 0
                               ; interrupts
setb  ET1                      ; enable timer 1
                               ; interrupts
```

Although covered in the following section, the serial Receive Data input line can also be used to generate an interrupt in the same manner. The serial port must be set into Mode 2 (a 9-bit UART with the baud rate determined by the oscillator). The setup would be

SM0 = 1

SM1 = 0

SM2 = 0

REN = 1

With these parameters set up, the first 1 to 0 transition of the RxD pin would create an interrupt.

Serial Ports

The 8051 family has a full duplex serial transmission circuit. The two pins P3.0 and P3.1 have the alternate function Receive Data (RxD) and Transmit Data (TxD). Full Duplex means that it can transmit a byte of data through the TxD output at the same time as it is receiving a byte of data at the RxD input.

The most popular serial protocol is the (in)famous RS232 and this is supported by the 8051 family. P3.0 is connected to a serial to parallel converter, while P3.1 is connected to a parallel to serial converter as in Figure 2.20. The rate at which data is shifted in or out is very important in RS232 transmission. One of the Counter/Timers is usually used to set the baud rate (i.e. bits/second of transmission). For more details on the RS232 standard see Chapter 5.

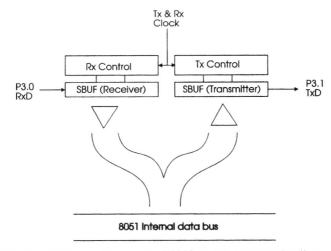

Figure 2.20 The 8051 configured for RS232 serial communications

Table 2.13

Input/Output Voltages Binary Logic	TTL/CMOS	Communication Voltages RS232
0	0 V	+10 V
1	+5 V	−10 V

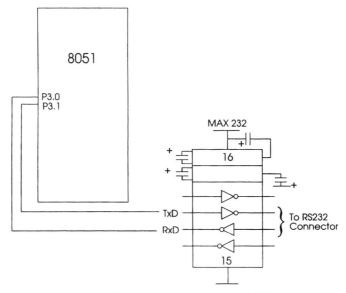

Figure 2.21 Changing the 8051's logic levels to suit RS232

Table 2.14

SM0	SCON.7	Serial Port control bit	Serial Port mode bit 0
SM1	SCON.6	Serial Port control bit	Serial Port mode bit 1
SM2	SCON.5	Serial Port control bit	Set by software to disable reception of frames for which bit 8 is zero
REN	SCON.4	Receiver Enable	1 = reception enabled
			0 = reception disabled
TB8	SCON.3	Transmit bit 8	(In Mode 2 or 3) is the 9th bit to be transmitted
RB8	SCON.2	Receive bit 8	(In Mode 2 or 3) is the 9th bit to be received
TI	SCON.1	Transmit Interrupt flag	1 = transmit buffer empty (cleared by software)
RI	SCON.0	Receive Interrupt flag	1 = byte received into receive buffer (cleared by software)

Table 2.15

SM0	SM1	Mode	Function	Baud Rate
0	0	0	Shift register I/O expansion	$f_{OSC}/12$
0	1	1	8-bit UART	variable
1	0	2	9-bit UART	$f_{OSC}/32$ or $f_{OSC}/64$
1	1	3	9-bit UART	variable

Figure 2.21 shows the MAX232 level converter connected to pins 10 (P3.0) and 11 (P3.1) to change the 8051 logic level signals into RS232-compatible voltage levels.

The serial port uses two registers and one of the timers. SCON is used for setting up and monitoring the serial port, while SBUF holds serial data. SBUF is physically *two* separate registers; one for holding data prior to serial transmission and one for accepting serial transmission and storing it before it is read by the CPU. Which register is accessed depends on whether the CPU performs a READ or WRITE to SBUF.

The control register SCON performs the functions listed in Table 2.14.

The flexibility of RS232 is the cause of the complexity of this register. It is capable of operating in 4 modes according to the value of SM0 and SM1 Table 2.15):

Mode 0:   The TxD pin outputs the shift register clock (at $f_{OSC}/12$) while the RxD receives and transmits the 8 bits of data. The CPU is 'told' to expect incoming data when RI is set to 0 and REN is set to 1.

Mode 1:   Data is handled in a conventional 1 start, 8 data, 1 stop RS232 style. The 10 bits are transmitted via TxD and received via RxD.

Mode 2:   This mode allows the use of parity since the 9th data bit is handled by TB8 or RB8. So in all, 11 bits are handled in the order 1 start, 8 data, RB8 (or TB8), 1 stop. It takes more effort to use since the programmer has to work out the parity of the data byte and then adjust the TB8 (or analyse the RB8) bit accordingly. This mode tends to be used for multiprocessor communication since it operates at $f_{OSC}/32$ or $f_{OSC}/64$ as set by the baud rate doubling bit PCON.7 (SMOD). (More of this in the following baud rate setting section.)

Mode 3:   Mode 3 is the same as Mode 2 except that the baud rate is variable.

Modes 1 and 3 are used for RS232 communication: Mode 1 for 1 + 8 + 1 and Mode 3 for 1 + 8 + P + 1.

Multiprocessor Communication

Modes 2 or 3 can be programmed to act as channels for intercommunication between two or more microprocessors. This was quite an innovation when it was introduced and, while the feature still has its uses, standardization of other serial multiprocessor protocols has rendered it somewhat in a niche market. Operation is as follows:

Making bit SM2 = 1 sets up the serial port so that it will only generate an interrupt if the 9th bit received (into RB8) is a '1'. Hence, if an 8051 is acting as a MASTER to several SLAVES, then it may only want to address one of them. It would set TB8 to a 1 and output the address of the slave processor to which it wanted to communicate. All slaves would have an interrupt generated, and so all would be programmed to respond by looking at the received byte to see if they were being addressed. Thereafter, only the addressed slave would listen and respond to the master. All the other slaves get on with their own programs until a byte was transmitted with the 9th bit as a '1'.

Baud Rate Generation

RS232 conventionally uses one of a limited range of bit rates. In most cases, it is generated from the system clock and uses one of the Timers. A common method is to use Timer 1 in its Auto Reload mode. To recap, this is a mode whereby TH1 is loaded into TL1 every time TL1's register overflows to 00. Before we detail some commonly used frequencies and their Timer load values, there is yet one more register to explain. This is PCON. PCON is the Power Control register, and the 8051 only uses 3 of its bits for internal functions and 2 bits as general-purpose flags which the programmer can utilize.

Table 2.16

SMOD	PCON.7	Double Baud Rate. When SMOD = 1 a double baud rate is generated if Timer 1 is used and serial mode 1, 2 or 3 is selected
GF1	PCON.3	General-purpose user flag
GF0	PCON.2	General-purpose user flag
PD	PCON.1	Power Down PD = 1 activates Power Down mode until the processor is reset. The clock is stopped and minimal power is consumed. Only data held in the on-chip RAM is preserved, but all other settings are lost
IDL	PCON.0	Idle. IDL = 1 disconnects the clock from the CPU but not from the Interrupt, Timer or Serial ports. All registers and settings are preserved. The processor can be woken up either by an interrupt or a full reset

Baud Rate Generation using Timer 1

Table 2.17

Baud Rate	Freq MHz	PCON.7 SMOD	Timer TH1	Baud Rate	Freq MHz	PCON.7 SMOD	Timer TH1
300	6	0	CC	1 200	16	0	DD
300	11.059	0	A0	2 400	11.059	0	F4
300	12	1	30	2 400	12	0	F3
300	16	0	75	2 400	16	1	DD
600	6	0	E6	4 800	11.059	0	FA
600	11.059	0	D0	4 800	12	1	F3
600	12	1	98	9 600	11.059	0	FD
600	16	0	BB	9 600	12	1	F9
1 200	6	0	F3	19 200	11.059	1	FD
1 200	11.059	1	D0	19 200	12	1	FD
1 200	12	0	E6				

The formula for calculating the TH1 value given oscillator frequency and baud rate is

$$TH1 = \{256 - (K \times \text{Oscillator Frequency}) \div (384 \times \text{Baud Rate})\}$$

where K = 1 for SMOD = 0 and K = 2 for SMOD = 1.

Mode 0 Example

```
; Mode 0: serial data exits and enters through the RxD pin.
; The TxD pin. The TxD pin outputs the shift clock. In
; mode 0, 8 bits are transmitted/received starting
; with the least significant bit. The baud rate is
; fixed to 1/12 the oscillator frequency.
;

  org    00h
  mov scon, #00h       ; set up for mode 0
loop:
  mov sbuf, #0aah      ; transmit aah
  jnb ti, $            ; wait until done
  clr ti               ; clear transmit flag
  jmp loop             ; repeat again
  end
```

Mode 1 Example

```
; This program transmits hex value 'aa'
; continuously across the serial port of an 8051 in
; mode 1
; It uses timer 1 at 1200 Baud

  org 00h
  mov scon, #01000000B ; set serial port for
                       ; mode 1 operation
  mov tmod, #20h       ; set timer 1 to auto reload
  mov th1, #0ddh       ; set reload value for
                       ; 1200 baud at 16MHz
  setb tr1             ; start timer 1
  clr ti
```

```
loop:
  mov sbuf, #0aah          ; transmit 'aa' hex out on
                           ; TxD line
  jnb ti, $                ; wait until done
  clr ti                   ; clear transmit flag ready
                           ; to go again
  jmp loop                 ; repeat again

  end
```

Mode 2 Example

```
; A program to transmit the hex value 'aa'
; continuously out of the serial port of in mode 2
; at 1/32 the oscillator frequency
;

  org 00h

  mov scon, #10000000Bh    ; set up for mode 2
  setb smod                ; baud rate equals 1/32
                           ; osc. Freq
  clr ti                   ; ready to transmit

loop:
  mov sbuf, #0aah          ; transmit aah
  jnb ti, $                ; wait until done
  clr ti                   ; clear transmit flag
  jmp loop                 ; repeat again

  end
```

Mode 1 Example

```
; This program continuously receives a byte
; entering the serial port pin RxD and puts the
; data out on port 1.
;

  org 00h

  jmp main
```

```
        org 23h                 ; starting address of
                                ; serial interrupt
        jmp serial_isr

main:
        mov scon, #50h          ; set up serial port for
                                ; mode 1 with receive
                                ; enabled
        mov tmod, #00100000B    ; set up timer 1 as
                                ; auto-reload 8-bit
                                ; timer
        mov th1, #0ddh          ; baud rate equals 2400
                                ; baud at 16MHz
        setb smod               ; set the double baud
                                ; rate bit
        mov ie, #10000001B

        setb tr1                ; start timer 1
        clr ri                  ; ensure receive
                                ; interrupt flag
                                ; is clear
loop:
        jmp loop                ; endless loop
                                ; (unless interrupt
                                ; occurs)
serial_isr:                     ; serial interrupt
                                ; service routine
        mov p1, sbuf            ; move the data to port
        clr ri                  ; clear the ri bit
        reti                    ; return to the main
                                ; program

        end
```

RS422 Serial Transmission

Although the most famous, RS232 is not the only serial data standard. One of the common industrial standards which has the advantage of faster, longer haul and more reliable data transmission is the RS422. This standard uses a twisted pair differential signal to achieve 60 m at up to 10 Mbaud. None of the 8051 family have this transmission standard built in, so an example is shown in Chapter 5.

The I²C Serial Bus

Another serial standard which is available in the family is the **I²C** or Inter Integrated Circuit bus. This two-wire system is used for interconnecting such devices as LCD drivers, remote I/O ports, RAM, EEPROM or data converters. Many variants of the

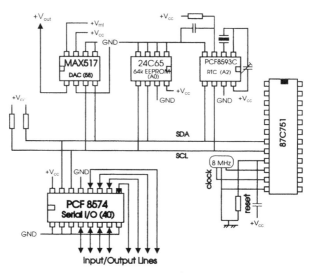

Figure 2.22   Hardware application of the I²C standard

range of the 8051 family are fitted with this standard. Figure 2.22 shows a simple hardware application of the standard. Chapter 5 gives more detail of the standard and the devices which can use it.

Related to I²C is the **CAN** serial standard which was originally developed for the automotive industry by Bosch. The 8XCE958 and 8XC952 both have the CAN standard interface protocol as part of their I/O capability.

Bit Addressing

There is an area in the DATA RAM where each bit can be individually addressed and manipulated. Intel highlight this as the family's 'Boolean Processing Capability' and many of the instructions operate on single bits. The 16 bytes from 20_H to $2F_H$ provide the 128-bit addressable locations as shown in Table 2.18.

The instruction mov C, 13 loads the carry flag with the contents of bit location 13 (0CH). This corresponds to bit 5 of DATA RAM BYTE 21 otherwise known as bit 21.5 Single-bit Boolean operations would then take place with other bits with instructions such as:

```
anl C, 52H      ; logical AND with bit 2 of RAM
                ; address 2A
orl C, 6CH      ; logical OR with bit 4 of RAM
                ; address 2D
mov P1.6, C     ; output bit to Port 1 bit 6
```

Table 2.18

Data RAM Byte	BIT							
	7	6	5	4	3	2	1	0
2F	7F	7E	7D	7C	7B	7A	79	78
2E	77	76	75	74	73	72	71	70
2D	6F	6E	6D	6C	6B	6A	69	68
2C	67	66	65	64	63	62	61	60
2B	5F	5E	5D	5C	5B	5A	59	58
2A	57	56	55	54	53	52	51	50
29	4F	4E	4D	4C	4B	4A	49	48
28	47	46	45	44	43	42	41	40
27	3F	3E	3D	3C	3B	3A	39	38
26	37	36	35	34	33	32	31	30
25	2F	2E	2D	2C	2B	2A	29	28
24	27	26	25	24	23	22	21	20
23	1F	1E	1D	1C	1B	1A	19	18
22	17	16	15	14	13	12	11	10
21	0F	0E	0D	0C	0B	0A	09	08
20	07	06	05	04	03	02	01	00

Of course, these bits could be given meaningful names so that the program could start with:

```
pump        equ    00
start       equ    01
emergency   equ    02
heat        equ    03
```

In this way, binary status parameters can be stored efficiently. A common sequence might have a structure such as

IF **heat** = ON AND **pump** = OFF AND **start** = ON
then set **emergency** flag

A bit contrived, perhaps, but easily implemented with

```
mov C,   heat
anl C,   start
cpl C
orl C,   pump
cpl C
mov emergency,  C
```

Only if **heat** = 1, **start** = 1 and **pump** = 0 would **emergency** contain a 1. There are many other ways of solving this – maybe with a test such as a `jc . . . jump if carry set` etc.

Using the ADC in the 80550 Family

The 80552 family is a functional derivative of the 8051 in that it has all the same facilities and more, namely:

- 8k × 8 EPROM (83552) expandable externally to 64 kbyte

- 256 × 8 RAM expandable to 64 kbyte

- 2 standard 16-bit Timer/Counters

- 1 additional 16-bit Timer/Counter coupled to 4 capture and 3 compare registers

- 8 multiplexed analogue inputs coupled to a 10-bit ADC

- 2 8-bit Pulse Width Modulated outputs

- 5 8-bit I/O ports (plus 1 8-bit port shared with the analogue inputs)

- 1 full duplex UART

- 1 I²C serial I/O with byte-oriented master/slave function

- 1 on-chip Watchdog Timer.

Available in a 68-pin PLCC, the functional layout is shown in Figure 2.23.

The larger amount of data RAM means that the 256 locations must be split:

- locations 0–127 are ordinary read/write data RAM, directly and indirectly addressable

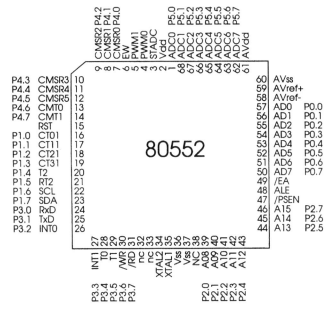

Figure 2.23 Functional layout of the 80552

- locations 128–255 are actually two separate sets of RAM:

  ◆ Performing a direct address Read or Write access to the SFR

  ◆ Performing an indirect address Read or Write access to the additional RAM space

```
mov r0, #90H
mov A, #90H        ; fetches SFR location 90H - i.e.
                   ; Port 1
mov A, @r0         ; fetches data RAM location 90H
```

Obviously with more functions, the SFR gets more complex as more of its 128 bytes are needed to control and report on the status of the microcontroller.

Analogue to Digital Conversion

The analogue to digital (ADC) input circuitry consists of a 10-bit converter fed by one of 8 multiplexed inputs (from Port 5). Part of the programming process involves selection of the appropriate channel. The sucessive approximation A–D conversion process takes 50 machine cycles, which is $20\,\mu S$ at a clock frequency of $12\,MHz$. The internal circuitry uses its own +5 V and 0 V supplies (AV_{DD} and AV_{SS}) which help to cut down on switching noise inherent in microprocessor circuits. The

Table 2.19

Address	Symbol	Function	Address	Symbol	Function
			Direct Hardware Byte Register		
FFH	T3	Timer 3	C5H	ADCON	ADC control
FEH	PWMP	PWM prescalar	C4H	P5	Port 5
FDH	PWM1	PWM register 1	C0H	P4	Port 4
FCH	PWM0	PWM register 0	B8H	IP0	Interrupt Priority 0
F8H	IP1	Interrupt Priority	B0H	P3	Port 3
F0H	B	B Register	AFH	CTL3	Capture Low 31
EFH	RTE	Reset/toggle enable	AEH	CTL2	Capture Low 2
EEH	STE	Set enable	ADH	CTL1	Capture Low 1
EDH	TM2H	Timer 2 High Byte	ACH	CTL0	Capture Low 0
ECH	TM2L	Timer 2 Low Byte	ABH	CML2	Compare Low 2
EBH	CTCON	Capture Control	AAH	CML1	Compare Low 1
EAH	TM2CON	Timer 2 Control	A9H	CML0	Compare Low 0
E8H	IEN1	Interrupt Enable 1	A8H	IEN0	Interrupt Enable 0
E0H	A or ACC	A Register	A0H	P2	Port 2
DBH	S1ADR	Serial 1 address	99H	S0BUF	Serial Data Buffer 0
DAH	S1DAT	Serial 1 data	98H	S0CON	Serial Control 0
D9H	S1STA	Serial 1 Status	90H	P1	Port 1
D8H	S1CON	Serial 1 Control	8DH	TH1	Timer 1 High Byte
D0H	PSW	Program Status Word	8CH	TH0	Timer 0 High Byte
CFH	CTH3	Capture High 3	8BH	TL1	Timer 1 Low Byte
CEH	CTH2	Capture High 2	8AH	TL0	Timer 0 Low Byte
CDH	CTH1	Capture High 1	89H	TMOD	Timer Mode
CCH	CTH0	Capture High 0	88H	TCON	Timer Control
CBH	CMH2	Compare High 2	87H	PCON	Power Control
CAH	CMH1	Compare High 1	83H	DPH	Data Pointer High Byte
C9H	CMH0	Compare High 0	82H	DPL	Data Pointer Low Byte
C8H	TM2IR	Timer 2 Interrupt reg	81H	SP	Stack Pointer
C6H	ADCH	ADC converter high flag	80H	P0	Port 0

reference supplies are also brought out separately on the chip, although many designers take the simplest route of connecting ref– to the analogue 0V (which in many cases is kept as the digital 0 V) and the ref+ to either the analogue +5 supply or a separate precision reference IC.

Good design principles dictate that:

- Digital and analogue grounds should be kept separate.

- The ADC reference voltage should be derived from a separate precision source.

- The analogue and digital supplies should be separately derived and controlled.

However, compromises sometimes must be reached and either through pragmatism, negligence or ignorance, you may find that because of switching noise creeping into the analogue electronics, the full 10-bit resolution is never achieved.

The result of the 10-bit conversion obviously needs two registers in which to store the result. The upper 8 bits are stored in the register ADCH and the lower 2 bits are stored in the ADC control register ADCON. Once the appropriate channel has been selected, conversion can be initiated by either a hardware signal or a software command. End of Conversion can be signalled either by an Interrupt or by polling the ADSC bit in ADCON. In more detail (Table 2.20):

Table 2.20

ADCON.0	ADDR0	Analogue Channel Select bit 0
ADCON.1	ADDR1	Analogue Channel Select bit 1
ADCON.2	ADDR2	Analogue Channel Select bit 2
ADCON.3	ADCS	ADC Start Convert. Set by software or external signal to start a conversion. It cannot be reset by software, only by End of Conversion
ADCON.4	ADCI	ADC interrupt flag: set when ADC ready to be read. It must be cleared by software. It cannot be set by software
ADCON.5	ADEX	Enable external Start of Conversion 0 Conversion cannot be started externally by STADC 1 Conversion can be started externally by STADC
ADCON.6	ADC.0	Bit 0 of ADC converted value
ADCON.7	ADC.1	Bit 1 of ADC converted value

Table 2.21

Source		Vector	Source		Vector
External 0	X0	0003	T2 capture 2	CT2	0043
Timer 0	T0	000B	T2 capture 3	CT3	004B
External 1	X1	0013	**ADC Complete**	**ADC**	**0053**
Timer 1	T1	001B	T2 compare 0	CM0	005B
Serial 0 (UART)	S0	0023	T2 compare 1	CM1	0063
Serial 1 (I²C)	S1	002B	T2 compare 2	CM2	006B
T2 capture 0	CT0	0033	T2 overflow	T2	0073
T2 capture 1	CT1	003B			

The most convenient way to use the converter is with interrupts. The full interrupt priority for this device is shown in Table 2.21:

The vector table at the head of the code listing should be amended to include

```
org  0053H
jmp  ADC_ISR
```

Thereafter, setting up code such as

```
mov ADCON, #00000110B        ; internal start select
                             ; channel 6
```

Conversion would be started with

```
orl ADCON, #00001000B
```

When the interrupt occurs, the first task is to reset the flag:

```
anl ADCON, #11101111B
```

Thereafter, the top 8 bits could be read for simplicity, or the full 10 bits could be used in whatever application is required.

If it is absolutely essential to **poll** the ADC rather than enable interrupts, then:

```
adc_wait:      mov a, adcon
               jnb acc.4, adc_wait
```

This illustrates the fact that it is impossible to perform a bit test on certain SFR registers. If the SFR register does not end in 0 or 8 then bit tests are unavailable – and ADCON is SFR register C5.

Timer 2

Like many of the more powerful MCS51 processor versions, the '552 boasts a Timer 2. This 16-bit timer has associated with it 4 × 16-bit capture registers, 3 × 16-bit compare registers and 9 interrupt sources.

Digital to Analogue Conversion (the PWM output)

A pulse width modulated (PWM) output is available from many microprocessors: both the 8051and PIC families have members which give this useful output. Once buffered by an external amplifier, a PWM output can be used as a form of digital to analogue converter (DAC). This is often used to drive external devices such as motors. Figure 2.24 shows the effect of varying the pulse width and the way in which the average voltage applied to the external device increases as the mark:space ratio increases.

Low   Half   High

Figure 2.24 The effect of varying pulse width

The '552 has two PWM output channels which generate pulses of programmable length and frequency. These are called PWM0 and PWM1. An 8-bit counter counts through the sequence 0–254 and the counter is continually compared with the contents of PWM0 and PWM1. Pulse Width Channel 0 is HIGH while the 8-bit counter is smaller than the PWM0 register. Pulse Width Channel 1 is HIGH while the 8-bit counter is smaller than the PWM1 register.

This means that if the PWM0 register is loaded with (say) 63 then the Channel 0 output will be HIGH while the Counter is lower than, or equal to, this value and LOW while the Counter is higher than this value. The averaged channel output is thus proportional to the contents of the corresponding PWM register. There is also a register PWMP which is a prescalar to set the repetition frequency of the output. The repetition frequency is given by:

$$f_{PWM} = f_{OSC}/(2 \times (1 + PWMP) \times 255)$$

By loading PWMP with different values, a repetition frequency range of 92 Hz to 23.5 kHz can be obtained with a 12 MHz clock. By loading the PWM registers with either 00 or FF, the PWM output channels will output a constant HIGH or LOW level respectively. There are no other controls for the PWM outputs and the output pins themselves are not used for any other purpose. Software operation is simple.

```
mov PWMP, #22          ; set output frequency for
                       ; 1 kHz 12 MHz clock
mov PWM0, #63          ; approx. 1:3 Mark-Space ratio
```

Alternatively, a ramp could be generated using code which incremented or decremented the PWM0 or PWM1 registers. Care would need to be taken that the register was not incremented at a rate faster than the pulse repetition frequency, so it is frequently implemented in a timer interrupt routine. A simple command such as inc PWM0 would have the desired effect.

The 80552 Watchdog Timer

The Watchdog Timer is enabled by connecting the /EW input pin (pin 6) to 0 V. It cannot be disabled by software. Timer T3 is used to implement the Watchdog Timer. It is an 8-bit timer (direct address FF_H) which, if allowed to overflow, will RESET the '552. The timer is incremented every $12 \times 2048 \times 1/f_{OSC}$ seconds, ≈ 2 ms with a 12 MHz clock.

The 8-bit timer T3 can be preloaded with numbers to give a watchdog interval of between 2 ms and 524 ms from a 12 MHz clock.

Bit 4 in the PCON control register is used to restart the Watchdog with the new value loaded into the T3 register. Hence the code

```
mov T3, #00
orl PCON, #00010000B
```

will set the maximum timeout from the Watchdog. This code needs to be repeated at intervals of not more than the 524 ms needed for a 12 MHz clock. In practice, the two lines of code would be embedded into a subroutine such as

```
wdt: mov T3, #00
     orl PCON, #00010000B
     ret
```

This would be tucked away with the subroutines in their usual position at the end of the program and could be called with an instruction such as

```
lcall wdt
```

If a shorter watchdog period is required, then (for example) mov T3, #206 would be incremented 50 times before overflow and hence would give a WDT period of 100 ms from a 12 MHz clock.

Parallel Expansion

A common I/O device used in the Intel families is the 8255 Programmable Peripheral Interface (PPI). This 40-pin IC has three 8-bit programmable I/O ports. The control can be seen to be fairly straightforward, as it requires:

● 8 data lines

● 2 address lines (to discriminate between the various output ports)

● 3 control lines /WR, /RD, and /CS

● 1 reset line.

Figure 2.25 shows how the MCS51 family could be interfaced. Note that it is treated as external DATA RAM in that information is both written to it and read from it. The decoder could be combinational logic or a PLD set up to select the 8255 whenever the correct address was output at the same time as a /RD (read) or /WR (write) signal.

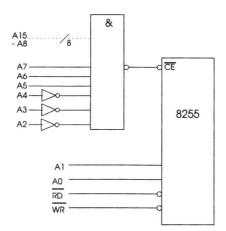

Figure 2.25 Interfacing the MCS51 family

This implies that it occupies a portion of DATA RAM – and the programmer must not forget which address it uses and inadvertently try to read/store data at that location.

An 8255 PPI requires 4×8-bit addresses:

● Port A Data

● Port B Data

Table 2.22

Port A	E0	11100000
Port B	E1	11100001
Port C	E2	11100010
Port D	E3	11100011

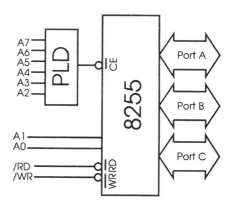

Figure 2.26   Logic circuit to decode addresses

- Port C Data

- Control – (setting up information for the port directions).

This is why there are 2 address lines required. To decode for addresses see Table 2.22.

Then a logic circuit similar to Figure 2.26 would need to be used.

The 8051 is unusual in having quasi-bidirectional ports, and as such they do not require to be given data direction information prior to their use. Like many other interface ICs (and microcontrollers count as that) the 8255 needs to be told which ports are to act as inputs and which as outputs. This information needs to be sent to the Control Port **prior** to any ports being used. Another old-style feature of the 8255 is that it is not *bit programmable*. This means that Port A must be wholly programmed as an output or an input port. The same applies for Port B. However, Port C can be nibble programmed, i.e. the upper nibble can be an output while the lower nibble can be an input.

The 8255 control byte should be

bit	7	6	5	4	3	2	1	0
function	Mode	Mode	Mode	A	Cu	Mode	B	Cl

Sending the binary word 10001010 to the control port would result in

Port A	Output
Port B	Input
Port C upper	Input
Port C lower	Output

The mode bits set the different handshake requirements but for straightforward uncontrolled nibble/byte transfer, Table 2.23 would suffice:

Table 2.23

7	6	5	A 4	CU 3	2	B 1	CL 0	A	B	CU	CL
I	O	O	O	O	O	O	O	O	O	O	O
I	O	O	O	O	O	O	I	O	O	O	I
I	O	O	O	O	O	I	O	O	O	I	O
I	O	O	O	O	O	I	I	O	O	I	I
I	O	O	O	I	O	O	O	O	I	O	O
I	O	O	O	I	O	O	I	O	I	O	I
I	O	O	O	I	O	I	O	O	I	I	O
I	O	O	O	I	O	I	I	O	I	I	I
I	O	O	I	O	O	O	O	I	O	O	O
I	O	O	I	O	O	O	I	I	O	O	I
I	O	O	I	O	O	I	O	I	O	I	O
I	O	O	I	O	O	I	I	I	O	I	I
I	O	O	I	I	O	O	O	I	I	O	O
I	O	O	I	I	O	O	I	I	I	O	I
I	O	O	I	I	O	I	O	I	I	I	O
I	O	O	I	I	O	I	I	I	I	I	I

To output binary data 01010101 on Port A as previously programmed, use the sequence:

```
mov R0, #EOH
mov A, #01010101B
movx @R0, A
```

Chapter 3

The PIC Microcontroller

Early PIC Variants

The PIC16 family of microcontrollers was developed by Microchip Technology Inc. in the late 1980s. They are all 8-bit microcontrollers with on-chip CPU, memory and I/O (Input/Output). The original base devices were:

- 16C54, 16C55, 16C56, 16C57

- 16C71

- 16C84.

However, this basic range has been supplemented by many and various versions which offer a comprehensive set of CPU, memory and I/O variants.

One very useful range of devices for embedded applications is the PIC12C5xx. This is a range of 8-pin Dual In Line ICs which offer all the same facilities as the PIC16C5x

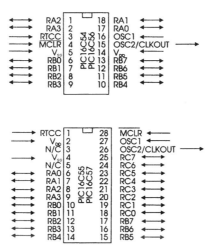

Figure 3.1 Pinouts of the 16C5x PICs

61

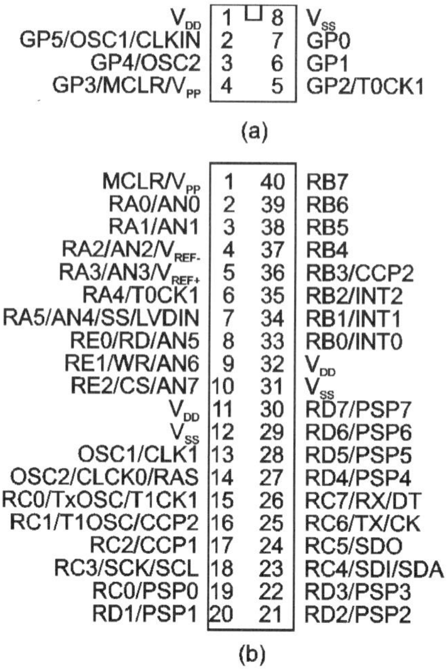

```
                  V_DD │ 1 ⊓ 8 │ V_SS
      GP5/OSC1/CLKIN │ 2    7 │ GP0
           GP4/OSC2 │ 3    6 │ GP1
       GP3/MCLR/V_PP │ 4    5 │ GP2/T0CK1
```

(a)

```
        MCLR/V_PP │ 1   40 │ RB7
         RA0/AN0 │ 2   39 │ RB6
         RA1/AN1 │ 3   38 │ RB5
     RA2/AN2/V_REF- │ 4   37 │ RB4
     RA3/AN3/V_REF+ │ 5   36 │ RB3/CCP2
        RA4/T0CK1 │ 6   35 │ RB2/INT2
   RA5/AN4/SS/LVDIN │ 7   34 │ RB1/INT1
       RE0/RD/AN5 │ 8   33 │ RB0/INT0
       RE1/WR/AN6 │ 9   32 │ V_DD
       RE2/CS/AN7 │10   31 │ V_SS
              V_DD │11   30 │ RD7/PSP7
              V_SS │12   29 │ RD6/PSP6
        OSC1/CLK1 │13   28 │ RD5/PSP5
     OSC2/CLCK0/RAS │14   27 │ RD4/PSP4
    RC0/TxOSC/T1CK1 │15   26 │ RC7/RX/DT
   RC1/T1OSC/CCP2 │16   25 │ RC6/TX/CK
         RC2/CCP1 │17   24 │ RC5/SDO
       RC3/SCK/SCL │18   23 │ RC4/SDI/SDA
         RC0/PSP0 │19   22 │ RD3/PSP3
         RD1/PSP1 │20   21 │ RD2/PSP2
```

(b)

Figure 3.2   Pinouts of (a) PIC12C5x; (b) PIC18Cxxx

but in a smaller package. There are many occasions when a small device is needed for switching, control or timing which does not need a huge amount of I/O lines (in this case, a maximum of 6), but is valued simply because it does not take up a large amount of board space or complexity of interfacing. For additional simplicity, it can be configured to operate from an internal RC oscillator which operates at approximately 4 MHz. Figure 3.2(a) shows the pinout of a PIC12C5xx.

At the other end of the scale, the 40-pin PIC18Cxxx is sold as a high performance RISC CPU with the capability of using a large program address space. Figure 3.2(b) shows the pinout of the PIC18Cxxx. As can be seen, and as is typical of complex, multifunction microcontrollers, the functions of the pins vary according to the application. This can sometimes lead to disappointment, since although the specification sheet lists an impressive array of capabilities of the device, they are not always available concurrently.

Briefly, all models feature:

- **Power on reset.** Under most operating conditions, the microcontroller will go to a known RESET condition without the use of external circuitry. However, for ultra-reliable reset operation, a capacitor can be connected to the MCLR (reset) pin.

- **Oscillator start-up timer.** Internal circuitry holds the microcontroller RESET for approximately 18 ms after the master clear is released.

Table 3.1

PIC Device Summaries						
Device	'C54	3prC55	'C56	'C57	'C71	'C84
Instructions	33	33	33	33	35	35
Max. Clock MHz	20	20	20	20	16	10
Instruction Width Bits	12	12	12	12	14	14
On-chip EPROM	512	512	1k	2k	1k	1k
General-purpose Registers	25	25	25	72	36	36
Hardware Registers	7	7	7	7	15	15
EEPROM Data Memory	–	–	–	–	–	64
Level of Hardware Stack	2	2	2	2	8	8
Interrupts	–	–	–	–	4	4
Input/Output Pins	12	20	12	20	13	13
Real Time Clock/Counter	1	1	1	1	1	1
A–D Converter	–	–	–	–	4	–

- **Watchdog timer.** A technique which, when used, allows the processor to escape from an endless loop fault condition if the watchdog timer is not regularly reset.

- **Security EPROM fuse for code protection.** When set, the security fuse prevents the internal program memory (the user program) from being read by external devices.

- **Power saving SLEEP mode.** A software command which shuts the processor down until it is RESET once more.

- **EPROM selectable oscillator options.** The microcontroller needs a frequency reference from which to operate; this could be chosen from RC components, quartz crystals or ceramic resonators. The oscillator section is hardware optimized to operate from 1 of 4 reference types. The following letters are the suffixes which are used:

Low cost RC oscillator	:RC
Standard crystal/resonator	:XT
High speed crystal/resonator	:HS
Power-saving low-frequency crystal	:LP

The CMOS semiconductor technology provides the following:

- Fully static design. If required, the clock can be stopped at any instant, and all data is held in the device memories. It is in effect frozen. Once the clock is restarted, it picks up from where it left off.

- Wide operating voltage range:

2.5–6.25 volts	16C5x
3.0–6.0 volts	16C71
2.0–6.0 volts	16C84
2.5–6.25 volts	12C5xx
2.5–5.5 volts	18Cxxx

- Low power consumption
  <2 mA @ 5 V, 4 MHz
  15 µA @ 3 V, 32 kHz
  approx. 1 µA standby @ 3 V

All memory for the PIC12C5xx and PIC16Cxx families is internal. In fact, it is difficult to extend the memory **externally**. The original concept of these microcontrollers was to be as self-contained as possible. The size of the EPROM can be seen from Table 3.1 to be fairly small. It is large enough for most tasks required of a small dedicated processor of this type. One unusual feature of the memory is the fact that the DATA memory (read/write – RAM) is 8 bits wide whilst the PROGRAM memory (read-only – OTPROM/EPROM/EEPROM) is 12 bits wide in the 12Cxxx and 16C5x; 14 bits in 16C71/16C84 and 16 bits wide in the 18Cxxx. The reason for the wide instruction word is that it not only holds the instruction, but also any immediate 8-bit data. So, for example, the instruction to load the W (working register) with the byte 59H has the opcode `movlw H' 59'`. This is encoded as the binary word $11000101 1001_B$. The first four bits 1100 represent the opcode, while 01011001 is, of course, the data word in binary. The data memory is completely separate from the program memory and also has a completely separate data bus structure to access it (technically it is called a Harvard Architecture).

As shown previously, the Intel 8031 family would need 2 bytes in 2 memory addresses to hold the instruction to load the ACCUMULATOR register with a number. The PIC12Cxxx/16Cxxx loads its equivalent – the Working register with 1×12-bit word in a single location. There are two consequences of this approach:

- The instruction set must necessarily be small to be contained in a few bits.

- If only one instruction fetch is required, then the machine must operate faster (i.e. in fewer machine cycles).

In fact there is another distinction between these two powerful industry standard processors – the basic 8051 uses 12 clock cycles to form a machine cycle, while the

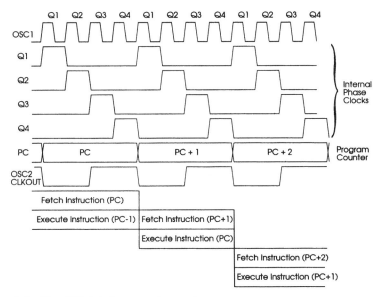

Figure 3.3 The PIC fetch execute cycle

PIC family uses only 4 clock pulses per machine cycle. Figure 3.3 shows how the PIC interleaves the act of fetching an instruction and executing it over the 4 clock cycles.

If timing diagrams leave you cold, then the simplest explanation is to point out that while one part of the processor is executing the current instruction, another part is fetching the next instruction.

Oscillator Connections

As mentioned, PIC microcontrollers will work with crystal oscillators, ceramic resonators or a simple RC circuit to produce these action-synchronizing clock pulses. Figure 3.4 shows the pinouts of various devices and a typical crystal

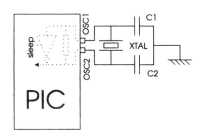

Figure 3.4 Pinouts and crystal oscillator connection

oscillator application. The values of capacitors C1 and C2 depend on how the PIC derives its oscillator.

A minimal component count can be achieved by using a 3-pin ceramic resonator. These are available in frequencies from approximately 4 MHz to 8 MHz but with a reduced accuracy of only some 0.5% (compared with a typical accuracy of some 0.002% for a quartz crystal). The resonator would be used as in Figure 3.5. When using the programmer, a resonator can be selected by opting for the same XT mode as in a crystal.

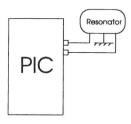

Figure 3.5   Resonator for PIC

Reset Circuit

When power is first applied to a microprocessor, it must be RESET so that the program can start from a known state. It will be RESET when the Master Clear (/MCLR) pin is connected to the 0 V supply. The PIC has internal circuits to perform this function at power-on and the simplest designs involve merely connecting the /MCLR pin directly to the +V supply or through a resistor to the +V supply. If, when the power supply is connected, the voltage rises too slowly, then this reset function may not work. In this case, Microchip recommend that the circuit of Figure 3.6 should be used. At switch-on, the capacitor is discharged. The PIC will be held reset until the voltage at /MCLR is above a threshold value. This will happen as the 100 nF capacitor is charged through the 470 W resistor.

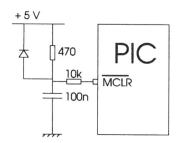

Figure 3.6   Reset circuitry for PIC

The PIC can be programmed to use an internal RC oscillator to generate a fixed 72 ms timeout on power-up. This is one of the options available when programming the IC. Other than that, there is also an Oscillator Start-up Timer which delays for 1024 oscillator cycles after the power-up delay. It ensures that the crystal or resonator has started and stabilized.

Internal Architecture

Instruction Set Summary

In the following descriptions and applications, the aim is primarily to explain the function of the 2C5xx or 16C5x PIC. Each instruction is a 12-bit word divided into an OPCODE which specifies the instruction type and one or more operands which further specify the operation of the instruction. (The more sophisticated members of the family use a 14-bit word.)

The 8-bit data bus connects two circuit components together:

● The register file composed of up to 80 addressable 8-bit registers. The first 8 of these are reserved for control or status reporting. The rest are general-purpose read/write registers. (Input/Output is 'memory mapped' which means that the I/O ports appear as an internal register. Inputting or Outputting data is as simple as reading or writing data to or from a memory location.)

● An 8-bit wide ALU.

The Arithmetic and Logic Unit (ALU)

The 8-bit wide ALU contains but one temporary working register (W register). It performs arithmetic and Boolean (logical) functions between data held in the W register and any file register. It also does single operand operations on either the W register or any file register.

Program Memory

Up to 512 words of 12-bit wide on-chip program memory can be directly addressed. Larger program memories can be addressed by selecting one of up to four available 'pages' with 512 words each.

Devices are available with either One Time Programmable (OTP) memory devices, or at greater expense, UV erasable memory devices. These are mostly used for development or pre-production models. Once the design is 'mature' then the OTP variants would be used.

Execution of the program instructions is controlled by the Program Counter (PC). At power-up or after a Master Clear, the PC is loaded with an address (1FFH for the PIC16C54). This is the address of the next instruction to be fetched. It is automatically incremented to point to the next valid address. Since the 16C54 can only hold 1FFH instructions, the PC automatically 'rolls-over' to 000. Jump (GOTO) instructions, Call instructions, Bit-Test-and-Skip instructions or computed addresses modify the contents of the PC. They are the only instructions which use 2 instruction cycles (8 clock pulses).

The stack itself is implemented differently from the 8051 (and most other microprocessors come to that). It is not part of the DATA or PROGRAM memory and the stack pointer cannot be manipulated (read to or written from) as in most other microcontrollers. The simpler PIC models have a limited STACK capability in that it is only **two** 12-bit addresses deep (12C5xx or 16C5x) or **eight** 14-bit addresses (16C6x, 16C7x, 16C8x). Care must be used to ensure that stack overflow or underflow does not occur by over-enthusiastic use of subroutines. This feature has sometimes been subjected to some criticism – mostly by engineers used to the tidy software discipline of multiple and nested subroutines. Consequently, PIC code tends to use a lot of **macros**.

The Stack

A stack is a device whereby an address can be saved for later reuse. It is most commonly seen when SUBROUTINES are CALLed. The sequence is:

1. The program code CALLs a subroutine.

2. The address of the instruction after the CALL is pushed onto the stack.

3. The program relocates to the subroutine by loading its address into the Program Counter.

4. When the subroutine has finished doing its business the RETURN instruction causes the address saved on the stack to be popped back into the PC.

5. Program execution continues where it left off before the CALL.

The Program Counter PC is a register which holds the address of the next instruction to be executed. The stack is usually referred to as a 'first-in-last-out memory', i.e. a RETURN statement will cause program flow to continue at the last address found on the stack.

	PC	stack1	stack2
If the PC contained 67_H	67		
After a CALL 103_H instruction	103	67	
After a return instruction	67		

Initially, the PC contained 67_H which means that the next instruction was located at that address. By executing CALL 103_H, the address 67 is loaded onto the stack while 103 is loaded into the PC. This means that the next instruction to be executed would be at location 103_H. At some point, the subroutine starting at 103_H would contain the instruction RETURN. This reloads the address stored at the current top of STACK (67_H) into the Programme Counter. Hence 67_H is the address of the next instruction to be executed.

Problems occur if the CALL instruction is made too many times:

PC contains 67H	67		
CALL 103H	103	67	
CALL 145H	145	103	67
CALL 150H	150	145	103

The situation is now that the original address can never be recovered no matter how many return instructions are issued. This is called stack overflow. If this becomes a limiting problem, the applications manual supplied with the PIC programmer gives a method of implementing a software stack.

12C5xx and 16C5x Register Structures

The registers can be divided into two categories:

● Operational Register Files (part of the DATA RAM)

● Special Purpose Registers (equivalent to the 8051 SFR).

Some of these registers use all of the byte or word to convey a particular value – such as the current count value in the Real Time Clock Counter (RTCC). Others rely on individual bits to control or monitor events – such as when bit 3 of the OPTION register controls whether the prescaler operates on the RTCC or Watchdog Timer WDT.

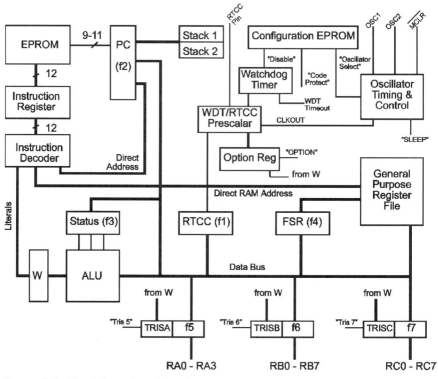

Figure 3.7 The internal register structure of the basic PIC

Operational Register Files

The lower 8 bytes of data memory are used for various register files (identified in Table 3.2 with an f-number) and as such have special functions to do with controlling or monitoring the microcontroller. They can all be read and written to by the software.

Special-purpose Registers

W	Working Register
TRISA	I/O Control Register for Port A
TRISB	I/O Control Register for Port B
TRISC	I/O Control Register for Port C
OPTION	Prescalar/RTCC Option Register

In more detail, the general and special registers are:

Address 00 (name f0). Indirect register addressing. This is not a physically implemented register. Addressing f0 calls for the contents of f4, the File Select Register (FSR), to be used to select a file register. f0 is useful as an indirect address pointer. For example,

```
ADDWF f0, W
```

will add the contents of the register pointed to by f0 to the contents of the W register and place the result in W. (If f4 = 0, the implication is that f4 is pointing to f0 itself. In this case, 00 will be read.)

Address 01 (name f1). Real Time Clock/Counter Register. This can be loaded and read by the program as any other register if required. In addition, its contents can be incremented by an external signal edge applied to the RTCC pin or by the internal instruction cycle clock. In this case, the instruction cycle clock is updated at a rate of (oscillator frequency)/4. An 8-bit prescalar can be assigned to the RTCC by writing the appropriate values to control bits in the OPTION register. So, if the PIC operates from an 8 MHz clock and the prescalar is set to 1:64 the RTCC register would be incremented every 32 μs.

Table 3.2

Address	Name	Function
00	f0	Indirection Register
01	f1	Real Time Clock/Counter Register
02	f2	Program Counter
03	f3	Status Word Register
04	f4	File Select Register
05	f5	Port A
06	f6	Port B
07	f7	Port C
08–0F	f08–f0F	'C54 'C55 'C56 General-purpose Register Files
	f10–f3F	'C57 Bank 0 Register Files
	f30–f3F	'C57 Bank 1 Register Files
	f50–f5F	'C57 Bank 2 Register Files
	f70–f7F	'C57 Bank 3 Register Files

Address 02 (name f2). Program Counter. The program counter generates the addresses for up to 2048 × 12 on chip EPROM locations containing the program instruction words.

Part	PC Width	Stack Width
'C54	9 bit	9 bit
'C55	9 bit	9 bit
'C56	10 bit	10 bit
'C57	11 bit	11 bit

The program counter is reset to all 1s and is auto-incremented with each instruction unless the result of that instruction changes the PC itself. Hence the first instruction should be placed at 11111111_B which is $1FF_H$, but many designers program a NOP (No-Operation) at this location so that the PC rolls over to the more conventional start of 000. When the 9-bit register contains `111111111`, incrementing this by 1 causes the PC register to load with `000000000`.

Address 03 (name f3). Status Word Register. This contains the arithmetic status of the ALU and CPU. The individual bits are set or reset as a result of the preceding operation. For example, if the W register contains $D0_H$ (208_D) and $A0_H$ (160_D) is added to it, the result should be 170_H (368_D). The W register is only 8 bits wide (i.e. only capable of holding FF_H (255_D) as its largest number), so after the addition, the register would actually contain 70_H (112_D) and the C bit would be set.

If the W register contains 31 and 31 is subtracted from it, then as a result of this sum, the zero Z bit would be set.

Nearly all programs rely on testing bits in the status register and jumping to one or other parts of the program depending on whether or not the bit is set. Table 3.3 shows the bits in the status register.

Table 3.3

Bit	Use	Meaning
0	C	Carry
1	DC	Digit Carry (half carry)
2	Z	Zero
3	PD	Power Down (0 = SLEEP, 1 = ACTIVE)
4	TO	Time-Out (0 = Watchdog timeout)
5	PA0	16C54/C55: 3 General-purpose user flags
6	PA1	16C55 PA0 = 1 of 2 EPROM page select
7	PA2	16C57 PA0/1 = 1 of 4 EPROM page select

Most software programmers create a header file with these in for ease of programming. For example, it is easier to remember the instruction:

```
BSF STATUS, PA0; Set the PA0 bit in the Status word
```

than

```
BSF 3, 5
```

The BSF mnemonic is the **B**it **S**et in **F**ile instruction. In this case, bit 5 of file register 3 is being set. The bits and file register would be labelled with EQU commands such as

```
STATUS EQU 3 PA0 EQU 5
```

Address 04 (name f4). File Select Register. If it is not used for indirect addressing, f4 can be used as a general-purpose register for holding any 8-bit data. As a File Select Register, bits 0–4 select one of the 32 file registers which will act as one of the instruction operands. This allows an amount of auto incrementing through the register files. In the 16C57, bits 5 and 6 additionally identify and select the current data memory bank. All the other bits are read as 1s.

Address 05 (name f5). Port A. A 4-bit I/O register. Only the low order bits are used. Bits 4–7 are read as 0s.

Address 06 (name f6). Port B. An 8-bit I/O register.

Address 07 (name f7). Port C. PIC 16C55/57: 8-bit I/O register. PIC16C54/56: General-purpose-register. Writing to and reading from these three port registers directly affect the state of the I/O pins.

W Working Register

This busy register corresponds to the accumulator of the 8051 and other micro-controllers. For example, the only way to load a register would be via the two-instruction sequence:

```
movlw h'7E'; move (load) a literal value into w
movwf 6    ; move the contents of w to register 6
           ; (i.e. Port B)
```

Input/Output Operations

TRISA/TRISB/TRISC

I/O control registers: The previous example showed how to output a value to a port. In this case, it was H'7E' (B'01111110'). Before this can be done, the port must be set up so that the individual pins act either as an input or an output. The I/O control register is loaded with the content of the W register by executing the TRIS instruction. A '0' puts the contents of file register f5, f6 or f7 onto the selected I/O pins. These registers are write only and are set to 1 on a RESET. In other words the default state of the I/O pins after a RESET is as inputs.

For example, to set all of the pins of port B as outputs (16C5x):

```
movlw  b'00000000'  ; move (load) the binary word
                     ; 00000000 into w
tris   6             ; 6 = Port_B data direction
                     ; (TRIS) register
```

(Although for the more complex PIC devices, this procedure takes a few more steps to access the data direction register.)

Thereafter, a byte could be output with a command such as:

```
movlw  b'01010101'
movwf  6
```

Of course, labels are more meaningful, so a program would be written more usefully as:

```
PortB   equ    6

        org    h'1ff'
        goto   start
        org    0
start   movlw  b'00000000'
        tris   PortB
        movlw  b'01010101'
        movwf  PortB
                :
                :
```

Lines 2, 3 and 4 are the result of the reset vector structure of the PIC family.

At a reset, the Program Counter (PC) is set to

16C54	1FF	16C57	7FF
16C55	1FF	16C71	000
16C56	3FF	16C84	000

This is the location of the RESET VECTOR and the first instruction of the program which will be executed after a RESET. In the 16C5× family, the RESET VECTOR is set at the top of memory, so the first instruction must be to relocate to another part of memory. This relocation is usually a GOTO 0 or NOP (no-operation) instruction.

The following example continually reads the lower four bits of Port A and copies them to Port B:

```
PortA      equ      5
PortB      equ      6
           org      h'1ff'
           nop
           org      0
           movlw    b'00000000'
           tris     PortB        ; Port B = all outputs
           movlw    b'11111111'
           tris     PortA        ; Port A = all inputs

Loop       movf     PortA, w     ; Fetch data from PortA
           movwf    PortB        ; Output data to PortB
           goto     Loop         ; Loop continuously
```

If Port A has been set to all inputs, then individual bits could be set or cleared using the BSF or BCF command:

```
bsf    PortA, 1 ; set bit 1 of port A
bcf    PortA, 3 ; clear bit 3 of port A
```

Testing individual bits is achieved with the bit test and skip instructions BTFSC and BTFSS. The 'skip' means that the next instruction will be missed out if the bit tested is Clear (BTFSC) or Set (BTFSS).

Questions

1. Set up a 16C54 with RA0–3 as inputs and RB0–3 as outputs. Copy the inputs on the RA ports to the outputs on the RB ports.

2. Repeat question 1 but make RB0–3 the inputs and RB4–7 the outputs. When a '1' appears on RB0, RB4 should follow. Similarly the pairs should link RB1 & RB5; RB2 & RB6; RB3 & RB7.

3. Write a program which sets Port B bits 0–3 as inputs and Port A bits 0–1 as outputs. Port A should show in binary how many of the inputs are HIGH.

Motor Control

A fairly trivial example of a 16C54 in use is as a latching push-button motor control. Figure 3.8 shows the circuit diagram of a conventional pair of ON-OFF push-buttons controlling a motor. For this example, the motor is mains operated, and so a relay is used to supply power. The inputs are a normally open contact pair for ON and a normally closed contact pair for OFF. This is the usual configuration for

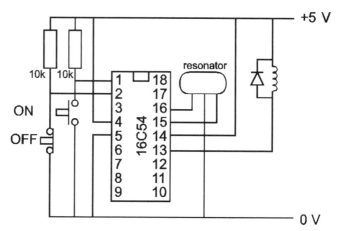

Figure 3.8 Motor controller

industrial controllers. Also note that there is no need to set up Port A since it is only used as an input port. This is the default condition for any port. Electrically, however, it is not wise to leave inputs 'floating'. Electrical noise could couple into the PIC, or electrostatic charges could damage the chip. The solution is either to 'pull-up' the unused inputs to the positive voltage supply through (say) a 10 kW resistor or to program unused pins as outputs. The following code should suffice:

```
; Motor controller
; 16C54
;
PORTA EQU        5
PORTB EQU        6                          ; bit 7 = motor
MOTOR EQU        7                          ; bit 2 = ON bit 3 = OFF
On               EQU        2
OFF              EQU        3
                 ORG        H'1FF'
                 GOTO       START
                 ORG        0
START MOVLW      B'11111110'
                 TRIS       PORTB            ; set bit 0 as output
                 BCF        PORTB, MOTOR     ; start with motor off
LOOP             BTFSS      PORTA, ON        ; is 'ON' pressed?
                 BSF        PORTB, MOTOR     ; if so turn motor ON
                 BTFSC      PORTA, OFF       ; is 'OFF' pressed?
                 BCF        PORTB, MOTOR     ; if so turn motor OFF

                 GOTO       LOOP

                 END
```

The only new instructions seen in this code example are the **BTFSS/BTFSC** Bit Test File and Skip instructions. These cause the program flow to Skip (the next instruction) if **B**it **S**et and Skip (the next instruction) if **B**it **C**lear. With the control buttons as shown:

'ON' is 1 (set) normally and 0 (clear) when pressed.

'OFF' is 0 (clear) normally and 1 (set) when pressed.

If the push buttons are in the normal position, then the next instruction to turn the motor on or off is skipped.

Counter Timer Register

OPTION. This defines the prescalar assignment (RTTC or WDT), prescalar value, signal source and signal edge for the RTCC. The OPTION register is write only and is 6 bits wide. By executing the OPTION instruction, the contents of the W register will be transferred to the option register. At RESET, the option register is set to all 1s.

This means that default condition is to use the Watchdog Timer at a rate of 1:128. If the Real Time Clock Counter is selected (bit 3 = 1), the other bits in the OPTION register allow you to choose whether the internal clock is used or whether an external clock is used. If the internal clock is selected, (bit 5 = 0), then the RTCC register (f01) will be incremented at a rate determined by the system clock and the

Table 3.4

OPTION REGISTER							
5	4	3	2	1	0		
RTS	RTE	PSA	PS2	PS1	PS0	RTCC RATE	WDT RATE
			0	0	0	1:2	1:1
			0	0	1	1:4	1:2
			0	1	0	1:8	1:4
			0	1	1	1:16	1:8
			1	0	0	1:32	1:16
			1	0	1	1:64	1:32
			1	1	0	1:128	1:64
			1	1	1	1:256	1:128
PSA prescalar assignment bit: 0 = RTCC 1 = WDT (Watchdog Timer)							
RTE signal edge: 0 = 0–1 1 = 1–0							
RTS signal source: 0 = internal clock 1 = EXTERNAL on RTCC pin							

PS bits of the OPTION register. If external is selected, (bit 5 = 1), then the choice is between whether the RTCC register (f01) is incremented when this external clock goes Low (1–0 transition; use bit 4 = 1) or when it goes High (0–1 transition; use bit 4 = 0).

For example, to set a 1:64 rate on the RTCC internal clock:

```
movlw b'000101'
option
```

The 16C5x Instruction Set

Notes

f   file register

d   destination: 0 = W register; 1 = file register

b   bit field

k   8- or 9-bit constant

SLEEP mode: when in a low power SLEEP mode, the IC is woken up either by a master reset (register information will be lost) or by the Watchdog timing out. This is a useful mode of operation for battery powered equipment since it takes less current from the supply when in sleep mode. The oscillator driver is turned off although the output ports maintain the same values that they had before the SLEEP instruction was issued. Current drain can be reduced further by setting the outputs to act as inputs while the PIC is asleep (if this is electrically acceptable in the application circuit!). There is a simple program and further explanation on page 107.

Simple Programming Examples

The PIC family has only one register – the W or working register. The rest of the registers are part of the inbuilt file registers (RAM). Like all other microprocessors, a lot of the user programs are concerned with transferring data between registers with such commands as:

ADDWF   Add working register to file register

ANDWF   AND working register with file register

IORWF   OR working register with file register

SUBWF   Subtract working register from file register

SWAPF   Swap upper and lower nibbles of file register

XORWF   Exclusive OR working register with file register

Table 3.5

Instruction	Flags	Operation
ADDWF f, d	C, DC, Z	Add W to f
ANDLW k	Z	W is ANDed with 8 bit k: result in W
ANDWF f, d	Z	AND W with f
BCF f, b		Bit b in register f is reset to 0
BSF f, b		Bit b in register f is set to 1
BTSFC f, b		If bit b in f = 0 skip the next instruction
BTFSS f, b		If bit b in f = 1 skip the next instruction
CALL k		Subroutine CALL
CLRF f, d		d = 0: f and W are zeroed; d = 1: f only zeroed
CLRW	Z	W register is cleared and Z flag set
CLRWDT	TO, PD	RESETS the Watchdog Timer and prescaler
COMF f, d	Z	Complements Register f
DECF f, d	Z	Decrements Register f
DECFSZ f, d		Decrements f, skips next instruction if zero
GOTO k		Unconditional Branch
INCF f, d	Z	Increments Register f
INCFSZ f, d	Z	Increments f, skips next instruction if zero
IORLW k	Z	ORs W with k; result in W
IORWF f, d	Z	ORs W with f
MOVF f, d	Z	Move register f to either W or back to itself
MOVLW k		Load k into W
MOVWF f		Move data from f to W
NOP		No Operation
OPTION		OPTION register is loaded from the W register
RETLW k		W is loaded with k and the PC from STACK
RLF f, d	C	f is rotated left through the carry
RRF f, d	C	f is rotated right through the carry
SLEEP	TO, PD	The processor is put into SLEEP mode
SUBWF f, d	C, DC, Z	2's complement subtract W from f
SWAPF f, d		Exchange upper and lower nibbles of f
TRIS f		The TRIS register is loaded with the W register
XORLW k	Z	W is XORed with k
XORWF f, d	Z	W is XORed with f

Each of these instructions has a destination register which is specified in the instruction itself. W and F are reserved constants (0 and 1 respectively) which inform the assembler of the destination register. For example:

```
SUBWF 12, W    ; W - F12 result placed in W
```

In practice, of course, file register 12 would be given a more meaningful definition in the EQUATES section, perhaps something like

```
MSB:   equ D'12'; D'' specifies decimal:
                  ; it could have been
                  ; B'00001010' O'16' or H'0C'
```

Then we could see a line such as

```
ADDWF MSB, F    ; W + F12 result in F12
```

The MPSTART software is distributed with a file called PICREG.EQU, reproduced on page 111. This is a full listing of all the register and useful function names. It may be used as a header file for a program by means of an instruction such as:

```
INCLUDE "PICREG.EQU"
```

The effect of this line is to act as though the code in the file named had been inserted at that point in the source code prior to assembly.

In a particular program, a file location has been EQUated to the word TEMP and that location is to be loaded with the binary number 10010110.

This could be done with the sequence:

```
MOVLW  B'10010110'
MOVWF TEMP
```

or even:

```
MOVLW value
MOVWF TEMP
```

if the constant **value** is equated to B'10010110'.

Like all other microprocessors, the PIC is capable of indirect addressing. It uses both register 0 and register 4. Reading or writing data to register 0 causes that data to be read or written to whichever register is held (pointed to) by register 4. For example:

```
MOVLW H'13'
MOVWF 4
MOVF 0, W
```

would actually cause the data in location 13 H to be loaded into the W register. An advantage of using indirection is that sequential memory locations can be manipulated easily by incrementing or decrementing the contents of the F4 register.

Members of the PIC16C5x family

The full membership of the PIC16C5x family are

- **16C54.** The base unit has already been described in the preceding pages.

- **16C55.** A variant of the 'C54 but is issued in a 28-pin Integrated Circuit (IC) and has an additional 8-bit I/O port called Port C. It has the same memory capacity, i.e.
  - ♦ 512 program memory locations
  - ♦ 32 data memory locations.

- **16C56.** An 18-pin IC with a 4-bit Port A and an 8-bit Port B (like the 'C54). It is an expanded version with 1024 program memory locations and 32 data memory locations.

- **16C57.** The same I/O capacity as the 'C55 (4-bit Port A, 8-bit Port B, 8-bit Port C). It has much larger memory capacity in that it has 2048 program memory locations and 80 data memory locations.

I/O Port Structures

Electrical Characteristics of the I/O Pins

Figure 3.9 shows the typical internal schematic arrangement of the I/O pins. All of the ports can be used for both input and output activities. Prior to use, the programmer must set the data direction on each active I/O pin. Only if the I/O control latch has a '0' written to it will the transistors Q1 or Q2 be able to be driven. This would happen if the programmer loads the W (working) register with a '0' (O = output) for each pin to act as an output.

For example, writing the byte 00001111 to the TRIS (the PIC name for a Data Direction Register) would cause its upper nibble to be an output while its lower nibble would be an input.

Any port (A, B or C) can source 40 mA total or sink 50 mA total. An individual I/O pin can source 20 mA and sink 25 mA. Thus pins could be used to directly drive output devices such as LEDs. However, the temptation to use this capability should be curbed when directly handling inductive or capacitive loads because of the risk of EMC problems. Most practical o/p devices should be driven with appropriate devices

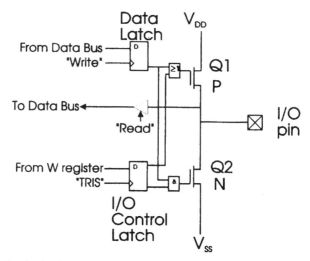

Figure 3.9 Equivalent circuit for a single I/O pin

– be it via discrete components such as transistors or via integrated circuit device drivers – Figure 3.10 shows a few possible devices. The Bipolar Transistor Darlington pair has a current gain of a minimum of 1000, so, for example, a DC load which draws 5 A would only draw a maximum of 5 mA from the I/O port. The MOSFET is a voltage operated device and draws very little current (mA) into the gate. The driver array is an IC with a set of Bipolar Transistor Darlington pairs.

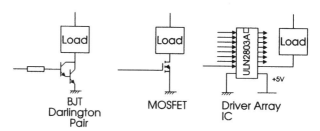

Figure 3.10 Various output devices

Omitted for simplicity are the freewheel (sometimes called back e.m.f.) diodes which are connected across the load with their cathodes to the positive supply. These protect the drive transistor against the very large e.m.f. which can be generated when turning off the current to an inductive load. Even with a 5 V supply, this reverse voltage could cause the collector (or drain) of the transistor to rise rapidly to 300 V or more. This is usually fatal for the transistor. The output driver IC has these diodes built in. The more sophisticated 16C64, 16C71, 16C74 and 16C84 offer the option of pull-up 'resistors' inside Port B so that input devices such as push-buttons, keypads etc. can

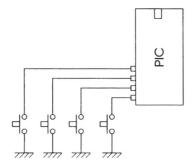

Figure 3.11 Direct drive using internal pull-up resistors

be directly connected to the PIC with no additional passive components. These resistors are actually formed from transistors and are referred to in the manuals as 'weak pull-ups'.

Timing

The PIC family all have a Real Time Clock Counter (RTCC) register (01). It can be read or written to just like any other register, but can also be incremented by either the system clock or an external signal (via the RTCC pin). The 16C5x family has no interrupt structure so the register must be regularly polled to see if the required time/count has elapsed. When the register reaches FF, it 'rolls over' to 00. Figure 3.12 shows the schematic arrangement of the RTCC. The bits RTE RTS PS2 PS1 AND

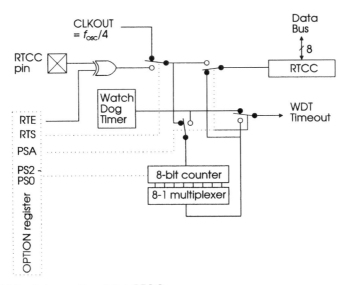

Figure 3.12 Schematic of the RTCC

PS0 are located in another control register called the OPTION register which is loaded from W by the OPTION command.

A 16C5x operating from a 4 MHz clock has an internal clock rate of 1 MHz ($foSc/4$). If the prescaler is set to 256, the RTCC register would be incremented every 256 µs. Thus the maximum time would be 256×255 µs or 39.68 ms.

Here is an example piece of code which is intended to produce a pulse train of 1 ms pulses with 1ms spaces. If the prescalar is set to 4, the RTCC is updated every 4 µs. If the RTCC is preloaded with 06 then after 250 increments, the RTCC will be at 0 (testable in the status register).

```
; Pulse Train Circuit
            ; 16C54              4 MHz Clock
RTCC        EQU 01
SIGNAL      EQU 00
PORTA       EQU 05
STATUS      EQU 03
Z           EQU 02
            MOVLW B'00000000'
            TRIS PORTA          ; all outputs
            BCF PORTA, SIGNAL
OUTER       MOVLW B'00000001'   ; divide by 4 (=4µS)
            OPTION
            MOVLW D'06'         ; 06 + 250 counts = 256
                                ; (same as 00)
            MOVWF RTCC
INNER       MOVF RTCC, W
            BTFSS STATUS,Z      ; test if done yet
            GOTO INNER          ; wait until time up
            MOVLW B'00000001'
            XORWF PORTA, W      ; complement signal
                                ; output
            MOVWF PORTA
            GOTO OUTER
```

Questions

1. Write a program for a 16C54 which outputs from pin RA1 a continuous stream of pulses ON for 1 ms and OFF for 5 ms.

2. A 12C508 has a push button connected to pin 7. When pressed, the output on pin 6 should produce 1 kHz tones for 0.2 s with 0.3 s between tone bursts.

3. A 12C508 has an input line on pin 2. Following a positive transition on this input, the output on pin 5 should go high for 1 s, followed by a 2 s pulse on pin 6 followed by a 4 s pulse on pin 7.

Table 3.6 Option

0	PS0	
1	PS1	
2	PS2	
3	PSA Prescalar Assignment bit	0 = RTCC 1 = WDT
4	RTE RTCC Signal Edge	0 increment on RTCC pin low to high 1 increment on RTCC pin high to low
5	RTS RTCC Signal Source	0 internal instruction cycle clock 1 transition on RTCC pin

Table 3.7

PS2	PS1	PS0	RTCC Prescalar
0	0	0	2
0	0	1	4
0	1	0	8
0	1	1	16
1	0	0	32
1	0	1	64
1	1	0	128
1	1	1	256

The Rest of the Family

This is a more difficult task than with the 8051. The PIC families are constantly evolving and new versions are coming out all the time. Table 3.8 shows the essential differences.

Many engineers start by considering the base members of the family – the 16C5x group – and after finding their limitations, move on to something more productive. In fact, technology has moved on so far from these base members, that they do appear to be extremely limited in what they can offer (except cost and simplicity, of course). The first real improvements came when Microchip started offering devices with interrupts and a better depth of stack. The drawback was that they then had to redesign the data memory map to accommodate all the extra facilities. The way that they did it was to create a direct memory map (Register Page 0) which was the default option and to have a second memory map (Register Page 1) which was

Table 3.8

Device	EPROM	RAM	I/O	Speed	ADC	Int	EEPROM	Serial	PWM
16C54	512	25	12	20 MHz					
16C55	512	25	20	20 MHz					
16C56	1 k	25	12	20 MHz					
16C57	2 k	25	20	20 MHz					
16C61	1 k	36	13	20 MHz		3			
16C62	2 k	128	22	20 MHz		10			
16C63	4 k	192	22	20 MHz		10		Y	2
16C64	2 k	128	33	20 MHz		8		Y	1
16C65	4 k	192	33	20 MHz		11		Y	2
16C71	1 k	36	13	20 MHz	4	4			
16C73	4 k	192	22	20 MHz	5	11		Y	2
16C74	4 k	192	33	20 MHz	8	12		Y	2
16C84	1 k	36	13	10 MHz		4	64		

accessed by setting one of the bits in the STATUS register (bit 5 – RP0). Most of the registers are merely duplicated, but some of them change function. The most common change is typically for the Register Page 0 registers to hold data, while the corresponding Register Page 1 registers hold the control/status information. This is the general idea, but some of the details vary from device to device, so care must be taken to study the data memory map carefully.

The following pages give some extra detail about the 16C71, 12C67x and 16C84.

PICs with ADC

The PIC16C71 of the PIC family has all of the attributes of the 16C5x with the following main differences:

- 4 extra software instructions

- 8 level deep hardware stack

- 4 8-bit analogue to digital converters

- 4 interrupt sources:

  ♦ external INT pin (RB0)

  ♦ RTCC timer

  ♦ A/D conversion complete

  ♦ interrupt on change of 4 Port B pins RB7–RB4.

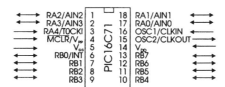

Figure 3.13   Pinout of the PIC16C71

The PIC12C671 and 672 are the 8 pin equivalent of the 16C71. Of the 6 I/O pins, up to 4 can be configured as ADC inputs. Once again, the main advantage is the compactness of the design.

The data memory map now becomes as shown in Table 3.9.

Table 3.9

00	Indirect	Indirect	80	Indirection register (mirrored)
01	RTCC	OPTION	81	Counter Timer + setup register
02	PCL	PCL	82	Program Counter low byte (mirrored)
03	STATUS	STATUS	83	Status Register (mirrored in both pages)
04	FSR	FSR	84	File Select Register (mirrored)
05	PORTA	TRISA	85	I/O port + data direction register
06	PORTB	TRISB	86	I/O port + data direction register
07			87	Available for general-purpose use
08	ADCON0	ADCON1	88	ADC control register
09	ADRES	ADRES	89	ADC result register (mirrored)
0A	PCLATH	PCLATH	8A	Program Counter high byte (mirrored)
0B	INTCON	INTCON	8B	Interrupt Control register (mirrored)
0C			8C	\
:			:	
:			:	
:			:	36 general-purpose registers (mirrored)
:			:	
:			:	
:			:	
2F			AF	/

The STATUS register also changes some of its functions, see Table 3.10.

Table 3.10

0	C	Carry
1	DC	Digit Carry (half carry)
2	Z	Zero
3	PD	Power Down: 0 = SLEEP, 1 = ACTIVE
4	TO	Time Out: 0 = Watchdog Timeout
5	RP0	Register Page 0: 0 = page 0 (00–7F), 1 = page 1 (80–FF)
6	RP1	Register Page 1 (not currently implemented)
7	IRP	Indirect Register Page: 0 = page 0 or 1, 1 = page 2 or 3 (not currently implemented). The aim of this control would be to indirectly access a different register page

OPTION Register

This has extra functions, as shown in Table 3.11.

Table 3.11

7	6	5	4	3	2	1	0		
/RPBU	INTEDG	RTS	RTE	PSA	PS2	PS1	PS0	RTCC	WDT
					0	0	0	1:2	1:1
					0	0	1	1:4	1:2
					0	1	0	1:8	1:4
					0	1	1	1:16	1:8
					1	0	0	1:32	1:16
					1	0	1	1:64	1:32
					1	1	0	1:128	1:64
					1	1	1	1:256	1:128

PSA	Prescalar assignment bit	0 = RTCC	1 = WDT	
RTE	RTCC signal edge	0 = L → H	1 = H → L	
RTS	RTCC signal source	0 = internal	1 = transition on RA4/RTCC pin	
INTEDG	INTerrupt edge select	0 = interrupt on falling edge	1 = interrupt on rising edge	

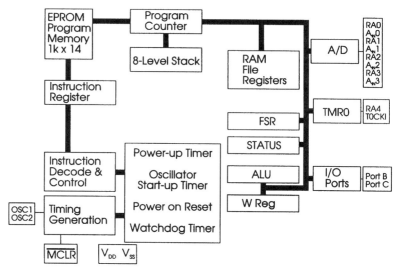

Figure 3.14 Internal structure of the PIC16C71

Instruction Set

Notes

f   file register

d   destination: 0 = W register, 1 = file register

b   bit field

k   8- or 9-bit constant

Table 3.12

Instruction		Flags	Operation
ADDLW	f,d	C, DC, Z	Add literal to W
ADDWF	f,d	C, DC, Z	Add W to f
ANDLW	k	Z	W is ANDed with 8-bit k: result in W
ANDWF	f,d	Z	AND W with f
BCF	f,b		Bit b in register f is reset to 0
BSF	f,b		Bit b in register f is set to 1
BTSFC	f,b		If bit b in f = 0 skip the next instruction
BTFSS	f,b		If bit b in f = 1 skip the next instruction
CALL	k		Subroutine CALL
CLRF	f,d		d = 0: f and W are zeroed; d = 1: f only zeroed
CLRW		Z	W register is cleared and Z flag set
CLRWDT		TO,PD	RESETS the Watchdog Timer and prescaler
COMF	f, d	Z	Complements Register f
DECF	f, d	Z	Decrements Register f
DECFSZ	f, d		Decrements f, skips next instruction if zero
GOTO	k		Unconditional Branch
INCF	f, d	Z	Increments Register f
INCFSZ	f, d	Z	Increments f, skips next instruction if zero
IORLW	k	Z	ORs W with k; result in W
IORWF	f, d	Z	ORs W with f
MOVF	f, d	Z	Move register f to either W or back to itself
MOVLW	k		Load k into W
MOVWF	f		Move data from f to W
NOP			No Operation
OPTION			OPTION register is loaded from the W register
RETFIE			Stack→PC, '1'→GIE
RETLW	k		W is loaded with k and the PC from STACK
RETURN			Stack→PC
RLF	f, d	C	f is rotated left through the carry
RRF	f, d	C	f is rotated right through the carry
SLEEP		TO, PD	The processor is put into SLEEP mode
SUBLW	k	C, DC, Z	Subtract literal from W
SUBWF	f, d	C, DC, Z	2's complement subtract W from f
SWAPF	f, d		Exchange upper and lower nibbles of f
TRIS	f		The TRIS register is loaded with the W register
XORLW	k	Z	W is XORed with k
XORWF	f, d	Z	W is XORed with f

Analogue to Digital Converter Programming

One of the main attractions of the 16C71 is the use of Port A as 4×8-bit analogue inputs with a typical 20 μs conversion time per channel. One of these four inputs can be selected via an internal sample and hold circuit and the 8-bit result of the conversion is left in register 09 – 'ADRES'. Setup and control of the ADCs is via the ADC control registers ADCON0 and ADCON1.

A similar difference exists in the 16C67x 8 pin DIL except that there is only one GPIO (General Purpose Input Output) port. Up to 4 pins of these ICs can be configured as ADC inputs. Like the 16C71, there is actually only one Analogue to Digital Converter. Software selects which of the inputs is routed to the ADC.

Port A is register 05 and its corresponding direction control register TRISA is mapped in page 1 of register file at address 85 H. Port A is a 5-bit wide port with pins RA0–RA4. Port pins RA0–RA3 are bidirectional whereas RA4 has an open collector output. Pins RA0–RA3 also have the alternate function AIN0–A1N3. The standard voltage reference for these pins is the supply voltage, although RA3 could further be used to input an external voltage reference (of NOT greater than the supply voltage). The A/D converter will thus provide a byte of information (0–255) which corresponds to a proportion of the supply voltage or whatever voltage is applied to pin RA3.

The two control registers ADCON0 (address 08h) and ADCON1 (address 88h) are used as shown in Table 3.13.

The ADC clock can be derived from the system clock or from an internal RC oscillator. The period of the RC oscillator varies, but is typically 4 μs–2 μs. Being a successive approximation conversion device, it takes 10 ADC clock periods to complete the conversion.

Table 3.13

ADCON0_7	ADSC1	ADC clock select
ADCON0_6	ADSC0	ADC clock select
ADCON0_5		General-purpose read/write bit
ADCON0_4	CHS1	Analogue channel select: 00 AIN0, 01 AIN1
ADCON0_3	CHS0	Analogue channel select: 10 AIN2, 11 AIN3
ADCON0_2	GO/!DONE	Set to 1 to begin conversion, reset to 0 by hardware to indicate conversion is done
ADCON0_1	ADIF	A/D conversion complete interrupt flag bit. Set when conversion is completed. Reset in software
ADCON0_0	ADON	0 = A/D converter module is shut off and consumes no operating current

Bits 4 and 3 of ADCON0 are used to select which of the inputs of Port A are to be selected for conversion:

ADCON0_7	ADCON0_6		ADCON0_4	ADCON0_3	
ADSC1	ADSC0		CHS1	CHS0	
0	0	$f_{OSC}/2$	0	0	AIN0
0	1	$f_{OSC}/8$	0	1	AIN1
1	0	$f_{OSC}/32$	1	0	AIN2
1	1	f_{RC}	1	1	AIN3

ADCON1 uses only two bits, which configure the function of pins RA0–RA3:

ADCON1_1	ADCON1_0	RA0	RA1	RA2	RA3	VREF
PCFG1	PCFG0					
0	0	A in	A in	A in	A in	Vsupply
0	1	A in	A in	A in	ref in	RA3
1	0	A in	A in	D in	D in	Vsupply
1	1	D in	D in	D in	D in	–

The 16C71 has 4 inputs which can be routed to a sample and hold circuit which is connected to an 8-bit successive approximation analogue to digital converter. The inputs are connected to Port A bits 0, 1, 2 and 3. If not used for ADC purposes, then any or all of these bits can be used as conventional digital I/O. Port A bit 3 can additionally be used to input a reference voltage other than the default value of 0 V. To provide all these various functions, additional control registers are required. These are:

ADCON0	08H	ADC control and status register
ADRES	09H	ADC result register (INTCON bit ADIE 0BH interrupt control register)
TRISA	85H	Port A data direction register
ADCON1	88H	ADC control register

The conventional address of PORTA	05_H	Port A pins when written
	05_H	Port A latch when written

A peculiarity of the 16C71 is that register files (RAM) exist in four blocks of 128 bytes. The blocks or pages are identified/set up by two bits (RP0 and RP1) in the status register (03). ADCON1 is mapped in PAGE 1. This means that in order to write to it you must SET bit 5 of the STATUS register (reviewed overleaf).

The following code is a continuous conversion of Channel 2:

```
;Continuous ADC on channel 2 pin 1 : output on Port_B
;16C71

PORTA    EQU 5
TRISA    EQU h'85'
PORTB    EQU 6
TRISB    EQU h'86'
ADCON0   EQU 8
ADCON1   EQU h'88'
ADRES    EQU 9
STATUS   EQU 3

  ORG 0

  BSF STATUS, 5         ; select page 1
  MOVLW b'00000000'     ; all outputs
  MOVWF TRISB
  MOVLW b'00000000'     ; all PortA's are analogue i/p
  MOVWF ADCON1
  BCF STATUS, 5         ; select page 0
  MOVLW b'11010001'     ; select: RC osc, ch2
  MOVWF ADCON0          ; turn on ADC

LOOP    CALL CONVERT
  MOVWF PORTB
  GOTO LOOP

CONVERT
  BSF ADCON 2           ; start ADC conversion
  WAIT BTSFC ADCON0, 2; is ADC over?
  GOTO WAIT             ; if not loop
  BCF ADCON0, 1
  MOVF ADRES, w
  RETURN

  END
```

The Port B data direction register is 86_H in this device instead of 06_H as in the 16C5x family. However, when selecting Page 1, the actual address becomes 06 – the result of having an addressing range of 127 bytes. Trying to write to 86_H will result in an assembler warning (when using the mpasm assembler).

The conversion must be to some purpose, so the following example is a simplistic temperature controller as in Figure 3.15. A temperature sensor is connected to AIN1.

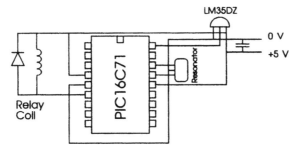

Figure 3.15 Simple temperature controller

If the temperature is below a digital value of 105, the heater (Port B bit 0) is turned on, while when the temperature rises above this level, the heater is turned off.

```
;Temperature controller
;16C71

PORTA     EQU 5
PORTB     EQU 6
MOTOR     EQU 0
TEMP      EQU 9
ADCON0    EQU 8
ADCON1    EQU 88H
STATUS    EQU 3

  BSF STATUS, 5          ; select page 1
  MOVLW B'00000010'
  MOVWF ADCON1
  BCF STATUS, 5          ; select page 0
  MOVLW B'11001001'      ; select: RC osc, CH1

LOOP
  MOVWF ADCON0           ; turn on ADC
  CALL CONVERT
  MOVWF TEMP
  MOVLW D'105'
  SUBWF TEMP, F
  BTFSC STATUS, 0        ; testing the Carry bit
  BSF PORTB, MOTOR
  GOTO LOOP
  BTFSS STATUS, 0        ; retest Carry bit
  BCF PORTB, MOTOR
  GOTO LOOP
```

```
CONVERT ; subroutine to return Temperature in W
  MOVLW -D'250'           ; enough for a 20 µs delay
  MOVWF TEMP

DLY DECFSZ TEMP           ; Decrement to zero
  GOTO DLY
  BSF ADCON0, 2           ; start ADC conversion
WAIT BTFSC ADCON0, 0      ; is ADC over?
  GOTO WAIT               ; if not LOOP
  MOVF TEMP, W
  RETURN
```

Status Register (03)

Several references have been made to the Status Register. This is readable and writeable and hold the results of the previous operation.

0. Carry/Borrow (C). Set if there is a carry in or out during add or subtract operations. It also holds the overflow bit during rotate operations.

1. Digit Carry/Borrow (DC). Used during BCD arithmetic and is set if there is a carry out from the 4th low order bit of the resultant.

2. Zero bit (Z). Set if the result of an arithmetic or logic operation is zero.

3. Power Down bit (PD). Set during power up or by a CLRWDT command. Reset by the SLEEP command.

4. Timer Out bit (TO). Set during power up and by the CLRWDT and SLEEP command. Reset by a watchdog timer timeout.

5. Page Select PA0.

6. Page Select PA1:

PA1	PA0	
0	0	Page 0 000–1FF
0	1	Page 1 200–3FF
1	0	Page 2 400–5FF
1	1	Page 3 600–7FF

7. Page Select PA2:

16C5x not used.

Interrupts on the PIC

The PIC 16C71 supports one of four interrupt sources:

External   External INT pin RB0. This is an edge sensitive input and can be set to trigger an interrupt on rising transitions of RB0 or falling transitions of RB0. The appropriate edge is set in the OPTION register.

Timer   Counter/Timer. When the RTCC Counter Timer0 register overflows from a count of 255 to a count of 0.

ADC   End of analogue to digital conversion.

Change   If one of RB7–RB4 changes state, then an interrupt can be triggered.

To assist this, there is a new register, the INTerrupt CONtrol register (INTCON) which is located at address $0B_H$. All pins of INTCON are readable and writeable: there are 5 controls which enable or disable interrupts and 3 flags which indicate when an interrupt has occurred. (The fourth flag, for ADC end of conversion, is located not in INTCON, but in ADCON0-bit 1.)

After a RESET (Master Clear) the GIE bit is cleared, so all interrupts are inhibited. If you wish to use interrupts in a program, this bit must be set along with the Enable bit of the interrupting source you want to use.

Table 3.14

INTCON7	GIE	Global Interrupt Enable. This is the master control which can be used to disable all interrupts: 0 = Disable all interrupts, 1 = Enable
INTCON6	ADIE	A/D conversion Interrupt Enable: 0 = Disable A/D interrupt, 1 = Enable A/D interrupt
INTCON5	TOIE	Timer TMR0 interrupt enable bit: 0 = Disable TMR0 interrupt, 1 = Enable TMR0 interrupt
INTCON4	INTE	INT interrupt enable bit: 0 = Disable INT interrupt, 1 = Enable INT interrupt
INTCON3	RBIE	RBIF interrupt enable bit: 0 = Disable RB change interrupt, 1 = Enable RB interrupt
INTCON2	TMR0	TMR0 overflow interrupt flag: 1 = RTCC overflow; reset by software
INTCON1	INT	INTerrupt flag: 1 = INTerrupt occurs; reset by software
INTCON0	RBIF	RB Port B change interrupt flag: 1 = one of Port B bits 7–4 changed; reset by software

When an interrupt occurs:

- GIE is cleared to prevent any further interrupt
- the return address is pushed onto the stack
- the PC is loaded with 004_H.

Once in the interrupt service routine, the source(s) of the interrupt can be determined by polling the interrupt flag bits (see software examples). The interrupt flag bit(s) must be cleared in software before re-enabling interrupts to avoid more interrupts. RETFIE, the return from interrupt instruction, exits the interrupt routine as well as sets the GIE bit to re-enable interrupts. (An alternative instruction 'RETURN' returns program control to the point at which the interrupt was invoked but does not re-enable interrupts.)

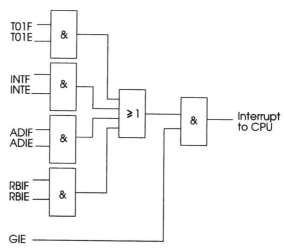

Figure 3.16 Interrupt structure of the 16C71

The following code sets up an interrupt and carries out an ADC conversion on channel 0 whenever the push-button connected to Port B pin 7 is pressed. For simplicity, some of the EQUates at the header of the program have been omitted.

```
        org    0
        goto   start
        org    4
        goto   ISR         ; Interrupt Service Routine

start   STATUS, 5           ; select page 1
        movlw  b'1000000'   ; Monitor port B bit 7
        movwf  PortB
```

```
movlw  b'000000000'  ; Select all port A as
                     ; analogue inputs
movwf  ADCON1
movlw  b'10000000'   ; Enable port B pull up
                     ; resistors
movwf  OPTION
bcf    STATUS, 5     ; reselect page 0
movlw  b'11000001'   ; RC osc, channel 0
movwf  ADCON0
clrf   INTCON        ; clear all interrupt
                     ; registers
bsf    INTCON, 3     ; enable Port B change
                     ; interrupts
bsf    INTCON, 7     ; enable interrupts
loop   :             ; main part of the program
       :             ; which waits for Port B
       :             ; changes, and then vectors to
       :             ; the interrupt routine
goto   loop

ISR  bsf ADCON0, 2   ; start new conversion
wait btfsc ADCON0, 2 ; test for end of conversion
  goto   wait
  movf   ADRES, w    ; get ADC data
  bsf    INTCON, 0   ; reset Port B interrupt flag
  retfie
```

This is a fairly simplistic program, kept simple by allowing only one type of interrupt – the Port B change interrupt. It would be perfectly possible to use interrupts triggered by the End of Conversion flag. In this case, there would have to be code at the start of the ISR to determine which routine caused the interrupt. Something along the lines of:

```
ISR  btfsc INTCON0, 0 ; Port B interrupt flag?
  goto   PB_ISR
  btfsc ADCON0, 1    ; ADC interrupt flag?
  goto   ADC_ISR
       :
       :
```

Questions

1. Write a program for a 16C71 which reads the analogue voltages on AD0 and AD1 and outputs the largest as a binary number on port B.

2. Write a program for a 16C71 which reads the analogue voltages on AD0, AD1 and AD2. If the values of AD1 falls between that of AD0 and AD2, make PB0 '1',

otherwise make PB0 '0'. This circuit function is also called a Window Detector.

3. Write a program which continuously reads the voltage on AD0. Output its largest value on Port B. If AD3 is '1' rather than '0', reset the value of Port B to the current vale input on AD0. This circuit function is called a Peak Detector.

4. Continuously sample the analogue voltage on AD0. When the value decreases, pin PB0 should pulse high for 1 ms.

Figure 3.17 shows one possible implementation of a push button operated a.c. power controller. The opto triac/triac combination is explained on page 117.

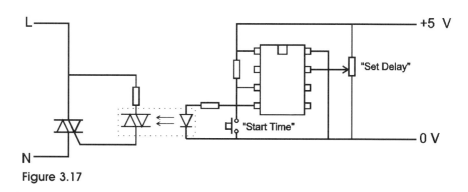

Figure 3.17

Stack

The device supports an 8-level stack. This is better than in the 16C5x family, but care must still be taken not to overflow this facility.

PIC16C84 with EEPROM Management

The main feature of this 16C5x-like device is that it uses EEPROM memory for both program and data. The program memory cannot be programmed 'in-circuit' or 'interactively', but it does provide a cheaper option than the EPROM version. When programmed with the MPSTART or other programmer, the contents of program memory will be erased before the new program is committed to EPROM. Many engineers use this IC as their main development tool prior to installation in OTP form – especially if they haven't yet committed themselves to an In Circuit Emulator – that essential workhorse of the professional developer.

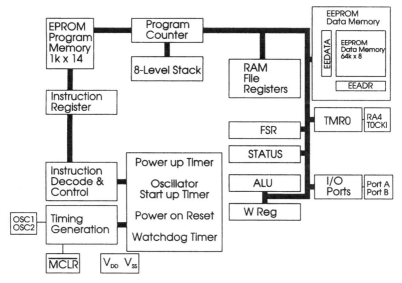

Figure 3.18 Internal structure of the PIC16C84

There are 36 data registers and an additional 64×8-bit EEPROM data memory locations. This means that essential data memory can be saved even though power is removed from the device. Figure 3.18 shows the internal structure of the PIC16C84.

It has the same 8-level stack and 4 interrupt sources as the 16C71, although the ADC interrupt is replaced by one related to the use of the EEPROM data memory structure.

The data register memory map for this device is shown in Tables 3.15 and 3.16.

Table 3.15

Page 0		Page 1	
00	indirect addr	indirect addr	80
01	RTCC	OPTION	81
02	PCL	PCL	82
03	STATUS	STATUS	83
04	FSR	FSR	84
05	PORTA	TRISA	85
06	PORTB	TRISB	86
07			87
08	EEDATA	EECON1	88 EEPROM data & control
09	EEADDR	EECON2	89 EEPROM address & control
0A	PCLATH	PCLATH	8A
0B	INTCON	INTCON	8B
0C	: general-purpose registers:		0C
0D	:		0D
:			:
:			:
2F			2F

Table 3.16 Status Register

0	C	Carry
1	DC	Digit Carry (half carry)
2	Z	Zero
3	PD	Power Down: 0 = Sleep, 1 = Active
4	TO	Time Out: 0 = Watchdog timed out, 1 = normal
5	RP0	Register Page 0: 0 = page 0, 1 = page 1
6	RP1	Register Page 1: not currently implemented
7	IRP	Indirect Register Page: not currently implemented

Option Register

This has a specific memory location in register page 1 (81).

Table 3.17

7	6	5	4	3	2	1	0		
/RPBU	INTEDG	RTS	RTE	PSA	PS2	PS1	PS0	RTCC	WDT
					0	0	0	1:2	1:1
					0	0	1	1:4	1:2
					0	1	0	1:8	1:4
					0	1	1	1:16	1:8
					1	0	0	1:32	1:16
					1	0	1	1:64	1:32
					1	1	0	1:128	1:64
					1	1	1	1:256	1:128
PSA	Prescalar assignment bit				0 = TOCKI		1 = WDT		
RTE	RTCC signal edge				0 = L → H		1 = H → L		
RTS	RTCC signal source				0 = internal		1 = transition on RA4/ TOCKI pin		
INT	INTerrupt edge select				0 = interrupt on falling edge		1 = interrupt on rising edge		
/PRBU	Port B pull-up enable				0 = pull-ups enabled		1= pull-ups disabled		

Status and Option registers have more or less the same function as in the 16C71, but the INTerrupt CONtrol register INTCON has slightly different functions to reflect the fact that the 16C84 has an EEPROM data memory rather than an ADC capability.

To implement the input push buttons as suggested in Figure 3.11, a '1' would have to be written to bit 7 of the OPTION register (81).

The INTCON register is defined as in Table 3.18

Table 3.18

INTCON7	GIE	Global Interrupt Enable. This is the master control which can be used to disable all interrupts: 0 = Disable all interrupts, 1 = Enable
INTCON6	EEIE	EEPROM write Interrupt Enable Bit: 0 = Disable EEIF interrupt, 1 = Enable EEIF interrupt
INTCON5	TOIE	Timer TMR0 interrupt enable bit: 0 = Disable TMR0 interrupt, 1 = Enable TMR0 interrupt
INTCON4	INTE	INT interrupt enable bit: 0 = Disable INT interrupt, 1 = Enable INT interrupt
INTCON3	RBIE	RBIF interrupt enable bit: 0 = Disable RB change interrupt, 1 = Enable RB interrupt
INTCON2	TMR0	TMR0 overflow interrupt flag: 1 = RTCC overflow; reset by software
INTCON1	INT	INTerrupt flag: 1 = INTerrupt occurs; reset by software
INTCON0	RBIF	RB port B change interrupt flag: 1 = one of Port B bits 7–4 changed; reset by software

Data EEPROM Management:

The 64×8 EEPROM data memory is accessed through two registers EEDATA (address 08) and EEADR (address 09). EEADR only has a valid range of 00–63. The storage and retrieval of data is controlled by the two registers EECON1 (address 88) and EECON2 (address 89) although EECON2 is not currently implemented. While it has a high number of write/erase cycles (1 000 000 typically), the write process is fairly lengthy, taking as it does 10 ms. Reading takes a conventional 1 clock cycle. EECON1 is as shown in Table 3.19.

Table 3.19

EECON1_7	U	Unused
EECON1_6	U	Unused
EECON1_5	U	Unused
EECON1_4	EEIF	Write completion interrupt flag: 1 = write complete – it must be reset by software
EECON1_3	WRERR	WRite ERRor: 1 = Write operation terminated abnormally – say by a WDT timeout or MCLR
EECON1_2	WREN	EEPROM write enable bit: 0 = EEPROM write disabled, 1 = EEPROM write enabled
EECON1_1	WR	Write control bit: 1 = initiate write cycle, automatically cleared by hardware on completion of write
EECON1_0	RD	Read control bit: 1 = initiate read cycle – automatically cleared by hardware on completion of read

There is a somewhat unusual sequence to follow to write data to the EEPROM. First, the WREN WRite ENable bit must be set, and then for each byte, the address and data is written to the EEADR and EEDATA registers respectively. Then there must be the sequence of:

● writing 55_H to the EECON2 register, followed by

● writing AA_H to the EECON2 register, followed by

● setting the WRite bit of the EECON1 register.

All interrupts must be disabled during this sequence.

Before the next byte can be written, the WR bit must be tested to see if the hardware has cleared it, or alternatively, an EEPROM interrupt handler can be used.

```
bsf     STATUS, 5       ; select register page 1
bcf     INTCON, GIE     ; disable interrupts
movlw   h'55'
movwf   EECON2
movlw   h'aa'
movwf   EECON2
bsf     EECON1, WR
bsf     INTCON, GIE     ; reenable interrupts
                        ; (if appropriate)
bcf     STATUS, 5       ; select register page 0
```

Reading the EEPROM data memory is somewhat more conventional. The programmer must:

● write the EEPROM address to the EEADR register

● set control bit RD in EECON1

```
bsf STATUS, 5           ; select register page 1
bsf EECON1, RD          ; set the READ bit
bcf STATUS, 5           ; select register page 0
```

The data will then be available in the EEDATA register on the next cycle. It remains in EEDATA until that register is overwritten by another EEPROM READ or WRITE action.

The Watchdog Timer

The watchdog timer uses an internal free-running RC oscillator which is either enabled or disabled as part of the programming procedure. Once the device has been programmed, it is not possible to 'turn off' an enabled watchdog, nor is it possible to 'turn on' a disabled watchdog. The watchdog will RESET the processor if the instruction CLRWDT is not regularly issued.

Like the RTCC, the WDT prescalar bits are set via the OPTION command/register to alter the nominal timeout period of 18 ms. The timeout periods vary with temperature, V_{DD} and process variations from part to part. The three bits PS2, PS1 and PS0 in the OPTION register determine the WDT rate prescalar (Table 3.20):

Table 3.20

PS2	PS1	PS0	Rate
0	0	0	1
0	0	1	2
0	1	0	4
0	1	1	8
1	0	0	16
1	0	1	32
1	1	0	64
1	1	1	128

The longest time period is when the three bits are set to 111. This represents a sleep period of some $128 \times 18\,\text{ms} = 2.304$ seconds. The following trivial example sends the PIC to sleep and allows it to reawaken after a timeout of $64 \times 18\,\text{ms} = 1.15\,\text{s}$. Upon reawakening, this trivial example merely increments the value output from Port B. Putting the PIC to SLEEP is a useful technique as it reduces power consumption to $<3\,\mu\text{A}$ (16C54).

```
              ; Watchdog timer and SLEEP command test
STATUS   equ 3
PG0      equ 5
TrisB    equ 6
PortB    equ 6
OPTI     equ 1

org      0

              bsf    STATUS, PG0    ; page 1
              movlw  B'00000000'
              movwf  TrisB          ; Port B = all output
              movlw  B'00001110'
              movwf  OPTI           ; select WDT @ 1/64 clock
              bcf    STATUS, PG0    ; page 0

              clrf   PortB
loop          sleep
              incf   PortB, f
              goto   loop

              end
```

Power Down Mode

The power down mode is entered by executing a SLEEP instruction. If enabled, the watchdog timer will be cleared but keeps running, the Power Down bit 'PD' in the status register (03) is cleared, the Time Out 'TO' bit is set and the oscillator driver is turned off. The I/O ports maintain the status they had before the SLEEP command was executed. This mode is used to draw minimal current, and it can be reduced further by all I/O pins at either V_{DD} or V_{SS} with no external circuitry drawing current from the I/O pin.

The device can be awakened by a watchdog timer timeout (if it is enabled) or an externally applied 'low' pulse on the RESET (Master Clear MCLR) pin. The PIC will stay in the RESET mode for one oscillator start-up timer period before normal program execution occurs. As explained previously, the 'PD' and 'TO' bits in the STATUS register indicate the cause of the system wake-up/reset.

TO	PD	
1	1	Power Up
0	X	Watchdog Timeout
1	0	SLEEP instruction
1	1	Clear Watchdog CLRWDT instruction

If the code uses the Watchdog and/or SLEEP facility, then it is important that the first few instructions establish the cause of the RESET. If the facilities are not used, then no such check would be needed.

If the PIC is awakened from a SLEEP by a WDT timeout then processing continues at the instruction following the SLEEP command.

The Prescalar can act on either the WDT or the Timer, but not both simultaneously. It may be necessary to reassign the Prescalar from one to the other under program control. This is fine, but it is **essential** to avoid a Timer or WDT timeout in the middle of the process. The way to avoid this is to issue a CLRWDT or CLRF TMR0 before reassigning that Prescalar.

Subroutines vs Macros

Subroutines operate in the PIC in the same manner as any other processor. The address of the Program Counter (PC) is pushed onto the stack and then the address of the subroutine is loaded into the PC. The routine re-enters the main routine by popping the return PC address off the stack and into the PC. Some of the code presented has already used subroutines.

The subroutines are usually kept together at the end of the program immediately before the end statement. They can be invoked with the instruction CALL, e.g. CALL CONVERT. They are identified by a label (in this case the subroutine is called CONVERT) and are terminated with some form of RETURN statement. This returns the program counter to the instruction after the CALL instruction. There are three versions of the RETURN instruction:

RETURN   Return to the instruction which follows the CALL instruction

RETLW k   As above, but load the W register with the number k first

RETFIE   As with RETURN, but enable interrupts as well.

The difficulty with subroutines is that they need a large stack capability to work effectively. Since most PIC models have a stack of depth 2 or 8, it can be a serious limitation to the programmer. It is the programmer's responsiblilty to ensure that subroutines are not nested too deeply. The effect of doing so has already been demonstrated – the return address is eventually lost, and the program would in all probability crash.

Macros

A macro is a piece of code which is caused to be inserted into the program whenever its name appears. It enables effective software reuse, and yet does not depend on the stack. A macro is written and included at the start of the program. Like the subroutine, it is given a label. When it is invoked (not by calling, but by just using the label name) the source code is assembled into the code at that point. This is one of its main disadvantages. If a subroutine is called on five occasions, there will just be five call statements relocating the program to the one instance of that subroutine. If a macro is invoked five times, all of the macro code will be assembled into machine code five times.

A simple example might be:

```
small_delay     MACRO                   ; 256 iterations
                local   sd
                movlw   0
                movwf   h'0A'
sd              decfsz  h'0A', f
                goto    sd
                ENDM
```

The macro is called small_delay and uses one local label sd. It loads the W register with 0 and decrements it until it reaches 0 once more. It uses register 0A to hold intermediate results. It is a very crude delay macro. In use, one might invoke the macro with a statement such as small_delay. At that point in the program all of the assembly language for the code would be included.

The following code has a similar action to the previous example of the SLEEP program. The value of Port B is incremented continuously. Well named MACROs make the code easier to read.

```
                :
                :
        clrf    PortB
  loop  small_delay
        incf    PortB, f
        goto    loop
                :
                :
```

The reason that macros are so useful and powerful is the fact that they can take parameters:

```
small_dly_2     MACRO    var1
                local    sd
                movlw    var1
                movwf    h'0A'
sd              decfsz   h'0A', f
                goto     sd
                ENDM
```

In this version, the macro can be invoked with a supplied parameter, e.g. `small_dly_2 55`. This would cause a delay count of 55 to be implemented, while `small_dly_2 87` would cause the delay to be 87 counts. In the first case, the variable var1 takes the value of 55, so when the line

```
movlw var1
```

is implemented, the assembler replaces the text 'var1' and replaces it with 55. The effect is to load the working register with 55. i.e. as if the line had been

```
movlw 55
```

The advantage is of course, that the value loaded into W can alter each time that `small_dly` is called. The number of macro parameters which can be handled is limited by the software assembler. They must have unique names such as var1, var2, var3 etc. and are usually separated with commas. Check with your documentation, but it is not uncommon for an upper limit of 32 parameters to be available. This is certainly more than the average programmer would ever want or dream of! The parameter need not just be number. On assembly, a simple text replacement process is worked through, so the parameter could be a label or register name. Some programmers have used the facility to extend the basic instruction set to the extent that it resembles a high level language rather than assembly language.

Questions

1. Write a MACRO to swap the contents of any two register files which are the specified parameters. e.g., the code

       swap H'0A, H'0B

   would cause registers 0A and 0B to have their contents swapped, while

       swap H'0C, H'0D

   would cause registers 0C and 0D to have their contents swapped.

   The definition of the MACRO would start

       swap MACRO var1, var2

2. Write a MACRO whose 1st line is

       ADC_in   MACRO var1

   which takes a parameter var1 of value 0, 1, 2 or 3 which results in the W register containing the data from the specified port.

3. Write a MACRO for a 16C84 which stores the contents of the W register in the EEPROM location specified by the passed parameter.

```
Equates file  "equates.h"


;
RTCC      equ 1 h
PC        equ 2 h
STATUS    equ 3 h        ; F3 Reg is STATUS Reg.
FSR       equ 4 h

Port_A    equ 5 h
Port_B    equ 6 h        ; I/O Port Assignments
Port_C    equ 7 h

ADRES     equ 9 h        ; 16C71 Special-purpose
                         ; registers
ADCON0    equ 8 h        ; 16C71 Special-purpose
                         ; registers
ADCON1    equ 88 h       ; 16C71 Special-purpose
                         ; registers
PCLATH    equ 0AH        ; 16C71 Special-purpose
                         ; registers
```

```
INTCON    equ 0BH      ; 16C71 Special-purpose
                       ; registers
TRISA     equ 85H      ; 16C71 Special-purpose
                       ; registers
TRISB     equ 86H      ; 16C71 Special-purpose
                       ; registers

;                      ; STATUS REG. Bits
C         equ 0 h
DC        equ 1 h
Z         equ 2 h
PD        equ 3 h
TO        equ 4 h
PA0       equ 5 h      ; 16C5X Status bits
PA1       equ 6 h      ; 16C5X Status bits
PA2       equ 7 h      ; 16C5X Status bits

RP0       equ 5 h      ; 16C71 Status bits
RP1       equ 6 h      ; 16C71 Status bits
IRP       equ 7 h      ; 16C71 Status bits
GIE       equ 7 h      ; 16C71 INTCON register bits
ADIE      equ 6 h      ; 16C71 INTCON register bits
RTIE      equ 5 h      ; 16C71 INTCON register bits
INTE      equ 4 h      ; 16C71 INTCON register bits
RBIE      equ 3 h      ; 16C71 INTCON register bits
RTIF      equ 2 h      ; 16C71 INTCON register bits
INTF      equ 1 h      ; 16C71 INTCON register bits
RBIF      equ 0        ; 16C71 INTCON register bits
ADCS1     equ 7 h      ; 16C71 ADCN0 register bits
ADCS0     equ 6 h      ; 16C71 ADCN0 register bits
CHS1      equ 4 h      ; 16C71 ADCN0 register bits
CHS0      equ 3 h      ; 16C71 ADCN0 register bits
GO        equ 2 h      ; 16C71 ADCN0 register bits
ADIF      equ 1 h      ; 16C71 ADCN0 register bits
ADON      equ 0        ; 16C71 ADCN0 register bits
PCFG1     equ 1 h      ; 16C71 ADCN1 register bits
PCFG0     equ 0        ; 16C71 ADCN1 register bits

LSB       equ 0 h
MSB       equ 7 h

TRUE      equ 1 h
FALSE     equ 0 h
```

Chapter 4

Interface Devices

This book is primarily concerned with embedded control. That is, a microprocessor-based system which is dedicated to a limited range of tasks, such as washing machine control, traffic light operation, credit card verification, etc. This chapter looks at the electrical characteristics of typical I/O devices.

Single Line Input

Switches come in two varieties: toggle (latching) and momentary (non-latching). Most microprocessor systems use momentary inputs. This is intentional so that upon a power failure, the system returns to a known state prior to the next power-up.

The switch consists of a pair of contacts, one of which has either a pull-up resistor or a pull-down resistor as illustrated in Figure 4.1. The signal is either normally-high switching low or normally low switching high. TTL does not respond very well to pull-down resistors, so most systems use a pull-up scheme regardless of the actual technology (i.e. CMOS or TTL).

```
; 8051
jnb P1.0, routine

; PIC
btfss PortB, 0
goto routine
```

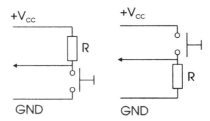

Figure 4.1 Passing contact switch circuits operates on the pull-down switch variant.

This is polled software which will cause either version to relocate to a piece of code with the label 'routine' if the button is being pushed when the input line is being sampled (= polled). An interrupt driven version is sometimes more appropriate if there are many possible inputs or the processor has several tasks to do.

The 8051 has vectored interrupts and provided that one of the two External Interrupt lines is used for the push-button, then the code will automatically relocate to the handling routine.

The External Interrupt vector is either location 0003 for X0 or 0013 for X1. These locations are specific to external interrupts and the location would normally contain a `jmp` instruction to access the Interrupt Service Routine.

The PIC16C5x family has no interrupts, and the others have no vectoring capability. Hence the code would be slightly more complicated while the location of the External Interrupt handler is determined.

The PIC relocates to address 04 when an interrupt occurs and interrupts have been enabled. If there is only one source of interrupt enabled, then location 04 could contain a simple GOTO to access the Interrupt Service Routine. If more than one interrupt source is enabled then the code at location 04 needs to ascertain which of the sources caused the interrupt before it **GO**s **TO** the appropriate service routine.

Both 8051 and PIC families have options for internal passive resistors on some I/O ports which will perform the pull-up function.

Optocouplers

One problem with switches is that as the contacts make, they tend to bounce under the force of the impact. This contact bounce can last for longer than 20 ms. The effect would be seen as a series of narrow pulses which occurred every time the switch was pressed. If it was part of a counting mechanism, then this would represent a serious source of error. Hardware solutions to the problem vary from discrete 'debouncers' which can be constructed from two logic gates to dedicated switch debouncing ICs. When the design includes a microcontroller, then a good alternative is to use software debouncing, i.e. to use software to discriminate between a bouncing input and frequent button pushes. The usual technique is to detect the first transition and then to wait for a short period before testing again.

Some inputs may come in the form of opto interrupters. The number of revolutions or angular velocity can be counted with the aid of a slotted rotary disc. Opto devices are inherently bounce free so the software interface is relatively simple to achieve, especially if software interrupts are to be used.

Sometimes it is important to detect the direction of rotation. If so, a quadrature detector made from two closely aligned opto switches, as in Figure 4.2, may be used.

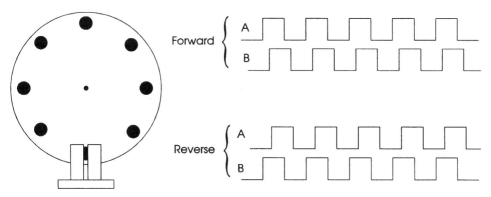

Figure 4.2 Quadrature detector using two opto switches and its output patterns

With this system, both speed and direction can be determined: speed by starting an internal Timer/Counter when the first hole is exposed and letting it run until the next hole is uncovered (allowing for a very slow/stopped disc, of course) and direction by comparing the relative transitions of the two signals as they arrive at the sensors.

Single Line Output

This is the most common output style, and can be used as a signal for anything from lighting LEDs (or incandescent lamps), operating motors, unlocking doors, turning on heaters – the list is endless. The basic concept is simple. The output line switches on a higher current device such as a relay, power transistor or triac to control the actual load.

As with any switching circuit, care must be taken when driving inductive loads. Sudden interruption of current through an inductor causes a very large EMF to be induced – the size of which is dependent on the switching speed, but is certainly large enough to destroy the switching device. The conventional solution to this problem is to use a diode in parallel with the load. This is the so-called 'flywheel' 'catching' or 'back-EMF' diode as in Figure 4.3. It prevents the voltage at the lower end from rising more than 0.7 V above the supply voltage. Use it.

```
; 8051
setb  P3.7        ; output high
clr   P3.7        ; output low

; PIC
bsf   PortB, 7    ; output high
bcf   PortB, 7    ; output low
```

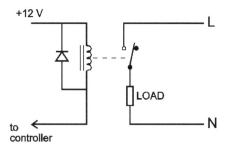

+12 V

L

LOAD

to
controller

N

Figure 4.3 A catching diode in parallel with an inductive load

Toggling (changing state) can be obtained with the following code:

```
; 8051
xrl P1, #00010000B

; PIC
movlw b'00010000'
xorwf PortB, f
```

It uses the fact that the exclusive-OR operation changes a bit state if the mask bit is 1, while leaving it unchanged if the mask bit is 0. In these examples, the bit 4 is to be toggled.

The truth table of an exclusive-OR gate is, of course:

A	B	X
0	0	0
0	1	1
1	0	1
1	1	0

When one of the bits is '1', the exclusive OR changes the state of the other bit. For more detail, see Appendix B.

To avoid difficulties with switching inductive loads, it is often worthwhile isolating the microcontroller from the load with an opto-isolator as was seen with the input signals. The load can operate from completely different power supplies than the microcontroller, offering both protection against switching transients and dangerous voltages. (There is still an issue of switching transients on the load side of the circuit. It is illegal to generate Radio Frequency Interference above certain levels and so care must still be taken. There are many cases where inadvertent interference has caused catastophic and tragic failures of nearby electrical/electronic equipment.) The simple opto triac circuit of Figure 4.4 can be used to drive some AC loads, but there is still the possibility of switching transients.

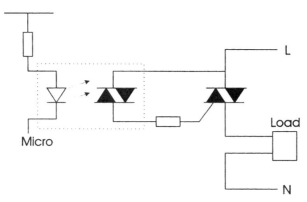

Figure 4.4 Use of an opto triac to drive an AC load

AC Control

A better solution is to use a zero-crossing detector which turns on the AC power when the mains supply is at its lowest. Separate integrated circuits are available for this, and if there is one spare input line on the microcontroller, then it is quite capable of detecting this event all by itself. The circuit of Figure 4.5 would suit.

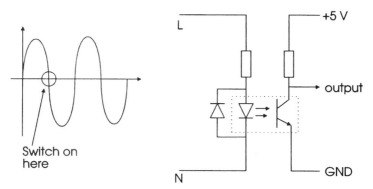

Figure 4.5 Zero-crossing detector for AC control

The reverse breakdown voltage of an LED is of the order of 5–10 V. Hence, when the supply reverses, the Junction Diode conducts and limits the reverse LED voltage to approximately 0.7 V. If a pulse is required on both zero crossing points (i.e. when the supply passes from negative to positive and positive to negative) then there would be TWO LEDs connected back-to-back. Care must be taken with the resistor power rating, since it has to drop a large voltage while passing the diode currents.

```
; 8051
; P1.7 = zero detect
; P1.0 = triac drive

    jb P1.7, next
    setb P1.0
    call Short_Delay
    clr P1.0

next:           :
                :


; PIC
; PortB, 7 = zero detect
; PortB, 0 = triac drive

    btfss PortB, 7
    goto next
    bsf PortB, 0
    Short_Delay
    bcf PortB, 0

next:           :
                :
```

In each case, a short delay is introduced to ensure that the triac is fully turned on. This will depend on the triac load. If the processor is involved with other tasks, then it would be more appropriate to use the zero crossing input to drive an interrupt routine as it may 'miss' the zero crossing pulse.

Burst Mode Control

A variation of this allows for 'burst mode' control to vary the amount of power delivered to a load. It is a technique normally reserved for heating loads where the time constants involved are relatively slow. The amount of power delivered is varied by the number of complete cycles of the mains supply in a fixed period. Figure 4.6 shows the principles.

Figure 4.6　Waveform of burst mode control

Phase Control

To control the amount of AC power supplied to a load the alternative technique of phase control is used. This can cause radio frequency interference problems, but is a common technique for dimming (incandescent) lamps and controlling the speed of universal motors. The microcontroller must sense the zero-crossing point and then apply a time delay before turning the triac on. Figure 4.7 shows typical waveforms.

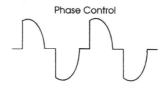
Phase Control

Figure 4.7 Waveform of phase control

```
; 8051
; P1.7 = zero detect
; P1.0 = triac drive

phase
  MACRO var1
  local loop
  local wait
  push r0
  mov r0, #var1
  mov TMOD, #10010001
  mov TCON, #00
loop mov TH0, #00
  mov TL0, #156 ; 100 μs
  setb TR0 ; start timer
wait cjne TL0, #0, wait
  djnz R0, loop
  setb P1.0
  call Small_Delay
  clr P1.0
  pop r0
  ENDM

; PIC
; PortB, 7 = zero detect
; PortB, 0 = triac drive
```

```
phase
  MACRO var1
  local loop
  local wait
  movlw var1
  movwf h'0A' ; use 0A
  movlw 0
  option
loop movlw h'156' ; 100 µS
  movwf RTCC
wait movf RTCC, w
  btfss STATUS, Z
  goto wait
  decfsz h'0A', f
  goto loop
  bsf PortB, 0
  Small_Delay
  bcf PortB, 0
  ENDM
```

Both macros take as a parameter the number of 100 µs units before the triac is to be turned on. (Of course, the timing depends on the clock chosen. It is assumed here that the 8051 is operating from a 12 MHz clock and the PIC from a 4 MHz clock.) The UK mains is at 50 Hz and the half-wave period is thus 10 ms. There would be 100×100 µs subdivisions of the half-wave waveform. This should be enough for all practical purposes of phase control. The useful range of parameters taken by the macro would be 1–99. This parameter is passed to a register which is decremented to control the number of 100 µs iterations.

Keypad Input

This is a cover-all name for a group of keys usually with a related function. Common sizes have 12 or 16 keys but there is no upper or lower limit. Section 1 showed that each key needs 1 data line to interpret it. Using 16 lines to indicate which of the keys on a keypad had been pressed is a wasteful use of resources. One improvement is called scanning and can be achieved in software, but is often left to a separate, dedicated IC. The procedure for scanning is as follows. Imagine the keys are arranged in a square array. Each (momentary) switch has two electrical contacts. One of these is connected to a common horizontal (ROW) connection and the other to a common vertical (COLUMN) connection. One at a time, each column line is changed from '0' to '1' and back to '0'. While it is high, the rows are examined, one at a time, to see if the '1' can be detected, and if so, on which line. By knowing which row and which column circuit was active at the time, the individual key pressed can be identified. This process is called multiplexing. Using software scanning, a 4×4 keypad would require 8 I/O lines as in Figure 2.10 in the 8051 section.

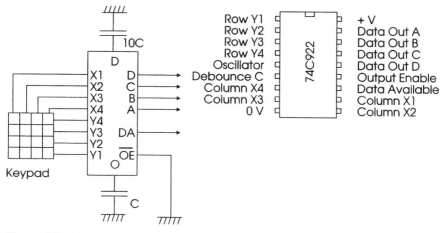

Figure 4.8 Use of a dedicated IC to interface to a keypad

Many applications opt for a dedicated IC such as the 74C922 in Figure 4.8. This would need only 5 lines connected to the microcontroller. These are called A, B, C, D and DA. A, B, C and D are binary data lines which indicate which key was pressed. DA is the Data Available line (active high) which indicates that a key is being pressed.

The advantages of the dedicated IC are that:

● Debouncing is carried out in the IC.

● The DA line is not active if two or more keys are pressed.

● Only 5 lines are required.

The binary code is not related to the indicated value on the key cap if a standard keypad is used. In this case, a software 'lookup table' must be used to convert returned code to intended value. One unfortunate characteristic of this IC is that the polarity of the DA pin is not suitable for driving the Interrupt line of the microcontroller. So, unless another logic device is available, or a simple 1-transistor inverter is used, the DA line must be polled to see if a valid key is being pressed.

Keyboard Input

Keyboards can have 102 keys or more, and the special encoder ICs usually work on an 8-bit data bus. These can be interfaced to microcontrollers relatively easily, but knowledge of the bus protocols and signal transfer levels is needed. These can be obtained from the manufacturers' data sheets. Alternatively, they can be treated exactly like a keypad encoder, but with a wider data field.

Seven-segment Displays

This is a common method of numerical display. They are invariably driven by binary (or BCD) to seven-segment display encoder ICs. Each display requires 4 bits. (LED displays can either be common anode or common cathode as in Figure 4.9. There is very little practical difference between them, other than the choice of the correct decoder.)

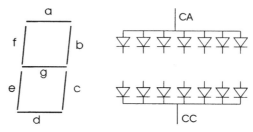

Figure 4.9   Seven-segment display unit

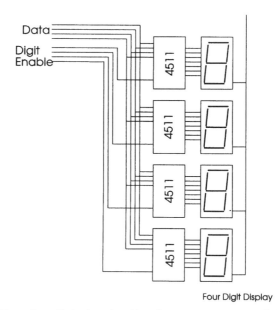

Four Digit Display

Figure 4.10   Use of multiplexing to drive four seven-segment displays

If several displays are required, then it would be inefficient to use four lines for each. A better structure would be to use a latching binary (BCD) to a seven-segment decoder driver such as a CMOS 4511 and a multiplexing technique, i.e. the four binary input lines would be connected to ALL the display drivers in parallel as in Figure 4.10. If there were to be (say) four displays then we would need four Latch Enable (LE) lines. To display the number '6789' on the display:

1. Output 6, operate LE for chip 1.

2. Output 7, operate LE for chip 2.

3. Output 8, operate LE for chip 3.

4. Output 9, operate LE for chip 4.

A four-digit seven-segment display could be driven in this way by eight output lines.

```
; 8051
; R7 MSB
; R6
; R5
; R4 LSB
; (data in lower nibble)

out_put  MACRO var
         setb   P1.var
         clr    P1.var
         ENDM
                  :
                  :


      :
      :
mov P1,   R4
out_put   4
mov P1,   R5
out_put   3
mov P1,   R6
out_put   2
mov P1,   R7
out_put   1
      :
      :

; PIC
; 0D MSB
; 0C
; 0B
; 0A LSB
; (data in lower nibble)

out_put  MACRO var
         movwf PortB
```

```
            bsf     PortB, var
            bcf     PortB, var
            ENDM
                      :
                      :


        :
        :
    movf    h'0A', w
    out_put 4
    movf    h'0B', w
    out_put 3
    movf    h'0C', w
    out_put 2
    movf    h'0D', w
    out_put 1
        :
        :
```

With larger tables, it would probably be more effective to use an auto-increment into a lookup area of RAM. However, with only four elements, there is not much to be gained.

The PIC is capable of sourcing up to 40 mA and sinking up to 50 mA. It is possible to use a single PIC to replace the circuit shown if high efficiency (low current) displays are used. The stobing/multiplex driving of the digits would need to be at a frequency so that the flicker would not be visible. 1–10 kHz is not uncommon in a technique which is often used to drive LED displays.

The attraction of this scheme is only that a PIC has relatively high current drive capability and is very cheap. If the display-PIC is to be driven by another microprocessor, then in all probability, a serial communication method might be more efficient. This would need to be a fast burst of data, otherwise the pause in the display would be noticable while the serial port is being serviced.

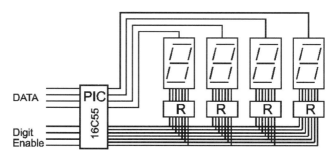

Figure 4.11

Liquid Crystal Displays

Numeric displays can be easily produced with seven segments. If full alphanumeric display is required, then the dot matrix layout is more appropriate. Each character is encoded in a rectangular region of dots. 8×5 or 10×8 are common formats as in Figure 4.12. Companies such as Hitachi produce dedicated ICs which will drive these displays and they are usually integrated with the LCD so that the designer has only to decide on the layout of the display. Typical formats are

● 8 characters × 1 line

● 16 characters × 2 lines

● 32 characters × 2 lines

● 32 characters × 4 lines.

The character sets can be standard or customized.

Graphical displays are also available if pictures rather than text are required, although the software drive for these would obviously be more complex.

The interface requirements vary, but for the simple text displays:

● 4 or 8 data lines (software selectable)

● 3 control lines (although a minimal write-only control can use just 2 control lines.

Chapter 6 shows in more detail the driving details of Liquid Crystal Modules.

Figure 4.12 Dot matrix display of the character 'h'

Analogue Input

Some microcontrollers have built-in ADCs – for example, the Intel 80552 and PIC16C71. If not provided internally, then a discrete IC must be used externally. Earlier ADCs presented their outputs in a parallel format and use 2 or 3 control or status lines. Most ADCs present their outputs in parallel and have 2 or 3 control or status lines. This is usually a variation on:

→	SC	start conversion
←	EOC	end of conversion
→	OE	output enable

More recent devices try to align the timing of the control signals with those of the microprocessor itself. So, for example, the ADC0803 can interface directly to many microprocessors; including the 8051 family. The only separate pin is the Chip Select CS which is derived from an address decode or I/O Port pin.

The ADC0803 8-bit Analogue to Digital Converter with Differential Inputs

The ADC0803 is a CMOS 8-bit successive-approximation analogue to digital converter. It is designed to operate from common microprocessor control buses with the three-state output latches driving the data bus. When dealing with micro-controllers, some of these control bus functions have to be provided explicitly by the I/O pins.

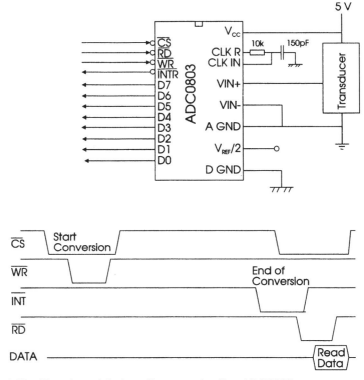

Figure 4.13 Pinout and timing diagrams for the ADC0803

The devices can operate with an external clock signal or, with an additional resistor and capacitor, using an on-chip clock generator.

If circuit board space is at a premium, and a microcontroller with inbuilt ADCs is not available, then a SERIAL ADC may be a solution. This device converts the analogue signal to digital and then outputs it one bit at a time in a serial fashion. The microcontroller controls the rate of output using a 'data clock' signal to the ADC.

The PIC family is not designed for external memory mapped devices, so the most convenient way to access an ADC is via a serial control medium. The alternative (parallel conversion) would take up the other valuable PIC resource of I/O lines. Serial conversion does seriously limit the rate at which conversion can take place, since it takes a finite time to pass the converted data across to the PIC. However, there are many applications for which the device is suitable.

TLC548 8-bit Analogue to Digital Converter with Serial Control

The TLC548 A/D peripheral integrated circuit is built around an 8-bit switched capacitor successive approximation ADC. It is designed for serial interface with a microprocessor or peripheral through a 3-state data output and an analogue input. It uses only the Input/Output clock (I/O clock) input along with the Chip Select (/CS) input for data control. The maximum clock input frequency of the TLC548 is guaranteed up to 2.048 MHz. Figure 4.14 shows the interface and timing diagram of the device.

Multiple Output ADCs

If more channels are needed, try something like the MAXIM MAX146 which offers 8×12 bit analogue inputs, and yet still uses a serial interface. The microprocessor not only has to start the conversion and retrieve the data, but also specify which of the channels it requires the data from.

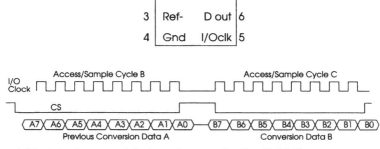

Figure 4.14 Interface and timing diagram for the TLC548

Analogue Output

The 'simple' method of achieving this in microcontrollers is via a PWM (pulse width modulated) output. This is available explicitly in some devices, or can easily be implemented with software. The idea is that the output is a pulse train with a mark:space ratio proportional to the desired analogue output, as shown in Figure 4.17. Passing the signal through a simple filter will reconstitute the signal as an analogue voltage. If a faster response is required, then a dedicated DAC might be considered necessary. This device will probably need 8 (or 10 or 12 or 14 or 16) bits of input as well as precision voltage references on the analogue side.

The MAX501 is a 12-bit, 4-quadrant voltage output multiplying digital to analogue converter. It is easily interfaced with microprocessors. Data is transferred into the input register in a right-justified 8 + 4 format. Table 4.1 defines the operations available.

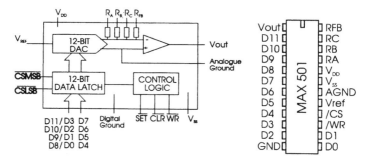

Figure 4.15 Schematic of the MAX501

Motor Drive Circuits

Stepper Motors

These are best understood by thinking of them as having four coils which must be sequentially energized to drag the rotor around as in Figure 4.16. The normal energization of the coils would be in the sequence A AB B BC C CD D DA. This is easy enough, but direct software control can be complicated by the need to accelerate and decelerate the rotor carefully to avoid 'rotor slip'. In practice, of course, there are many more steps to one revolution for a stepper motor. Common step values are 0.9°, 1.8° and 3.6°.

If the electrical interface has been set up to be coil A = bit 0, coil B = bit 1, coil C = bit 2 and coil D = bit 3, then using the bit output techniques already described, the software should output the data in the sequence:

Table 4.1

WR	CSMSB	CSLSB	LDAC	CLR	Set	Operation
X	X	X	X	X	0	DAC reg overridden by 1s
X	X	X	X	0	1	DAC reg overridden by 0s
0	0	1	1	1	1	Load MSB nibble into Input Register
0	1	0	1	1	1	Load LSB byte into Input Register
X	X	X	0	1	1	Transfer Input Register to DAC Register
1	X	X	1	1	1	No Operation
0	1	1	1	1	1	No Operation
0	R	1	1	1	1	Latching MSB nibble into Input Register (R = rising edge)
R	0	1	1	1	1	Latching MSB nibble into Input Register
0	1	R	1	1	1	Latching LSB nibble into Input Register
R	1	0	1	1	1	Latching LSB nibble into Input Register

```
00000001        00000100
00000011        00001100
00000010        00001000
00000110        00001001
```

Between each step there should be a delay which gives the motor mechanics time to react to the output drive from the microcontroller.

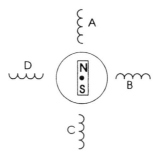

Figure 4.16 Stepper motor showing the four coils

DC Unidirectional

A simple relay or transistor can be used to turn the motor on and off. Speed control can be obtained via PWM (pulse width modulation) as already discussed and shown in Figure 4.17.

Figure 4.17 Pulse width modulation

DC Bidirectional (reversible)

The 'standard' way of producing this is with an 'H' bridge as in Figure 4.18. One direction is obtained by turning on devices A and D. The opposite direction is obtained by turning on devices B and C. This is common topology for servo mechanisms.

The electronics can be discrete devices or a special purpose IC. These are easier to use and often have suppression and level shifting electronics built in.

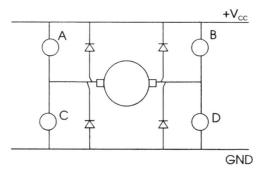

Figure 4.18 An H-bridge circuit for reversible motor control

AC Motors

Most AC motors rotate at a speed determined by the frequency of the electricity supply. With software, it is relatively easy to produce a variable frequency square wave as in Figure 4.19. Unfortunately, the square waves have a large harmonic content and this can cause heating and EMC difficulties. The efficiency is improved and the harmonic content decreased if a pseudo-sine wave is generated. Figure 4.20

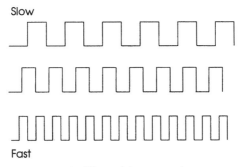

Figure 4.19 Square waves at different frequencies

shows a simplification of a pseudo sinewave. In practice, a real wave of this form would take many pulses to make the transistion from maximum positive to maximum negative. In software terms, this is quite fiddly to generate. The pulse repetition is fixed at a frequency above the normal human hearing limits. The pulse width depends on the amplitude of the sine wave at that point in the cycle. The most sophisticated type of control compensates for the fact that as the psuedo sine wave frequency gets lower, the magnetic circuit of the motor to be controlled can go into saturation – an undesirable situation. This is solved by reducing the apparent amplitude of the psuedo sine wave, i.e. all the pulse widths are reduced by a constant percentage so the filtered (averaged) sine wave output has a lower peak value. This would have to be tailored to each size/rating of motor to be controlled.

Figure 4.20 Pseudo sine wave

Questions

1. Design a software based switch debounce in both 8051 and PIC assembly language so that a single (bouncing) press-and-release of a momentary push button latches an output ON, while a second (bouncing) press-and-release latches the same output OFF. (As described on page 113, when a switch makes, the contacts can bounce for up to 20 ms; leading to false counts in digital circuitry.)

2. Design a solution the quadrature detector on page 115 which has two inputs for the optocouplers and two outputs – FORWARD and REVERSE. The software for 8051/PIC should detect

   ● if the wheel is stopped, in which case neither of the F or R outputs should be '1';

   ● the direction of rotation, in which case the appropriate output should be '1'.

3. Write a software 8051/PIC MACRO which takes a parameter in the range 0–10. This parameter is to control the percentage of power delivered in a burst mode heating scheme. Assume that a zero-crossing pulse is available at one of the inputs and the output should be a 2 ms pulse from an output immediately after the zero crossing input if the software determines that an output is appropriate. A parameter of 7 should output 14 outputs (i.e. 7 positive and 7 negative half cycles, followed by a non-output for the next 3 cycles (6 half cycles)).

4. Re-write the phase control MACROs for the US 60 Hz supply frequency.

5. It is a curious fact that the dedicated keypad switch IC 74C922 is more expensive than an unprogrammed 16C54. Write a PIC program to emulate the operation of this device.

6. Write a program for the PIC/8051 which has as its inputs a square wave clock and a single bit direction control and four coil drivers as outputs. Design a stepper motor driver. (An advance of this program would add the authenticity that a step increase in input frequency is not instantly matched by an output change, but by the output ramping up to the new frequency.)

7. Write a PIC/8051 MACRO to take an 8-bit value as an input parameter and to output a 1 kHz pulse stream, the pulse width of each is proportional to the input value.

Chapter 5

RS232 and Other Serial Standards

RS232

This is a standard which defines serial transfer of data from one point to another. The interface is intended for short cables of up to 15 m between the two devices and at data rates from 50 to 76800 baud (bits per second). It is a flexible, user customizable standard, and its inherent flexiblity is one of the causes of people's fear and mistrust of it. However, don't dispair. With a basic understanding of the alternatives and common configurations, it is relatively easy to quickly set up a working communications link.

The facts:

- It is also known as CCITT V24 standard.

- The standard is divided in four parts:

    ◆ definition of the plug, socket and pin assignment

    ◆ electrical signal characteristics

    ◆ functional description of the signals

    ◆ list of standard subsets of the interface signals for specific applications.

- RS232 divides the types of transmitter/receivers into either

    ◆ DTE Data Terminal Equipment

    ◆ DCE Data Communications Equipment.

This reflects the origins of RS232 from a teletypewriter system. Nowadays the terms usually refer to a computer system as DTE and a modem (for example) as DCE.

133

RS232 Connector Details

The connector system defined in the standard is the 25-way D-type connector, although there is also a 9-pin version which takes advantage of the fact that usually not all 25 pins are used. In fact, most RS232 applications use only nine of the pins and, if necessary, you can get away with using just three.

The pinouts are shown in Figure 5.1 and the pin function is described in Table 5.1.

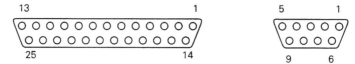

Figure 5.1 Pinouts of RS232 connectors

The asterisks show the signals which must be present for bidirectional data transfer. I have seen unidirectional data transfer systems which conform to a cut-down variant of RS232 with only 2 wires – TD or RD and SG.

Signal Details

Grounds

FG   Frame Ground is often left unconnected. It is connected to the chassis or metal frame of the equipments. It links the two cases and ensures that there

Table 5.1

25 Pin Connector	Signal Name	Abbreviation	9-pin Connector
1	Frame Ground	FG	–
2	Transmit Data	TD*	3
3	Receive Data	RD*	2
4	Request to Send	RTS	7
5	Clear to Send	CTS	8
6	Data Set Ready	DSR	6
7	Signal Ground	SG*	5
8	Data Carrier Detect	DCD	1
20	Data Terminal Ready	DTR	4

are no dangerous voltages on the equipment cases. It provides an element of protection against faulty equipment and can also have a noise-reducing effect.

SG   All of the RS232 signals are referenced to this ground. It is the one essential connection in the interface.

Data

TD   Data is transmitted from the DTE (computer) pin 2 to the DCE (modem) pin 2.

RD   Data is received at the DTE pin 3 from the DCE pin 3.

Handshaking (data readiness lines)

RTS   This signal is asserted when the DTE is ready to send some data.

CTS   The DCE will respond by asserting this line if it is ready to accept some data.

Handshaking (equipment readiness lines)

DTR   When asserted, this means that the DTE is powered up and ready.

DSR   This is the DCE equivalent to the DTR. It indicates that the DCE is operating and ready.

Modem Readiness

DCD   This is the DCE (modem) indicating to the DTE that a remote connection has been made.

Interface Cable Varieties

In the conventional computer to modem link (DTE–DCE) the connection is just made point to point:

```
1 ——————————————— 1
2 ——————————————— 2
3 ——————————————— 3
4 ——————————————— 4
5 ——————————————— 5
6 ——————————————— 6
7 ——————————————— 7
8 ——————————————— 8
20 ——————————————— 20
```

Figure 5.2 Conventional computer to modem link

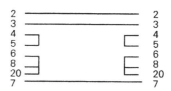

Figure 5.3 Loopback on RTS/CTS and DTR/DSR

Figure 5.4 Crossover connections

However, if the user wants to cut down on the number of lines, then each piece of equipment must be fooled into thinking that the other is in place and ready to go. (The obvious problem is that if for some reason one of the pair is not in place. Not to worry, this can be handled with software handshaking – more later.) When the DTE issues a Request to Send, it expects to get a Clear to Send signal. So, the obvious way to do this is to *loop back* the RTS into the CTS pin. Figure 5.3 shows the loopback present on both RTS/CTS and DTR/DSR.

Quite often, a designer may wish to use the RS232 protocol to interconnect two DTE equipments, such as two computers or microprocessors. In this case, *null modem* or *crossover* connections must be used, as shown in Figure 5.4.

Electrical Signal Characteristics

The RS232 standard uses negative logic and bipolar power supplies, i.e. a logic '0' lies between +3 and +25 volts, while a logic '1' lies between −3 and −25 volts. Figure 5.5 shows the voltage levels as the letter 'Q' is transmitted. 'Q' has the ASCII value of 51_H and RS232 specifies that bits are transmitted Least Significant Bit first. So,

$51_H = 01010001B$ – transmitted as 10001010

Figure 5.5 shows clearly the rest state of the line as a 'mark' or '1'. The transmitted byte is framed with a Start and a Stop bit; Start being a logic '0' and Stop being logic '1'. (Actually the Stop bit can be of length 1, 1.5 or 2 bit lengths long – as long as the sender and receiver agree beforehand which protocol is to be used, then all is well.)

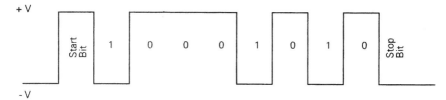

Figure 5.5 Voltage levels in RS232 transmission of the letter 'Q'

The next pitfall for RS232 users is that the standard allows for data to be transmitted as 5, 6, 7 or 8 bits, with or without odd or even parity and at a variety of baud rates. This is explained as follows.

Parity is a very simple method of error detection. If parity is enabled, it is declared to be ODD or EVEN and consists of an extra bit between the data bits and the stop bit. If the transmitter and receiver agree to use ODD parity, then the total number of data plus parity bits must add up to an odd number. If EVEN parity is decided upon, then the total number of data plus parity bits must add up to an EVEN number. If NO PARITY is opted for, then the parity bit is omitted, for example:

- 'J' transmitted in 7-bit even parity would be 10010010

- 'J' transmitted in 7-bit odd parity would be 10010011

- 'K' transmitted in 7-bit even parity would be 01010011

- 'K' transmitted in 7-bit odd parity would be 01010010.

('J' = 49H = 1001001 in 7-bit binary transmitted LSB first. 'K' = 4AH = 1001010 in 7-bit binary transmitted LSB first.)

The baud rate is simply the number of bits per second. Typically, these could be 75, 150, 300, 600, 1200, 2400, 4800, 9600, 19200 etc. This is not the rate at which data is transmitted, however. A 300 baud 8-bit system still requires the start and stop bits, so for each byte of data, 10 bits have to be transmitted. Hence the data rate would be 30 (300/(1 + 8 + 1)) bytes per second.

The rate should also be agreed before transmission commences. However, some equipment is auto-sensing in that it will try to sort out from the first characters transmitted the rate, parity and data size. It helps (or in some cases is essential) if the receiver knows what character to expect first. The space (20_H) is commonly used for this purpose.

All these variables – connection, DTE/DCE, rate, parity, word size – give lots of factors which can stop one device from transmitting to or receiving from another. Despite this, RS232 is still probably the most commonly used serial protocol in the world.

Even the simplest of the 8031 family has an inbuilt capability to handle Serial RS232 data. Example code for the family was shown on page 47 in Chapter 2.

Voltage Level Conversion

One of the difficulties of RS232 from a hardware designer's point of view used to be the awkward voltages used to characterize a '0' and '1'. Fortunately, this is eased by families of ICs which operate from a single +5 V power supply and yet convert the 0/+5 logic signal to and from an RS232 voltage level. Internally these devices have two power supply sections:

● a voltage doubler to raise the +5 volts up to +10 V

● a voltage inverter to convert the +10 V to a –10 V.

One of the first companies to exploit this technology was MAXIM, and Figure 5.6 shows two typical applications using RS232 voltage converters.

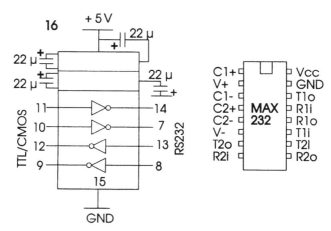

Figure 5.6 Using the MAX 232 RS232 voltage converter

Both 8051 and PIC microprocessor families have members which will output RS232 directly, but there are occasions when it is necessary to use the pins of a parallel input/output port. Before deciding to embark on this route, it is as well to consider what tradeoffs you are going to have to live with. In the first place, it is perfectly possible to design a pin output full duplex RS232 interface on a non-interrupt low capability microcontroller. However, it does take up a lot of memory space. Assuming that the IC is going to do something else useful as well as RS232, then it is as well to set some limits.

Full Duplex. Do you really need it? The difficulty is that in a micro like the PIC16C5x family which only has but one timer, if you are transmitting a byte, and an incoming byte is detected on the input pin, then the RTCC timer has to cope with timing the outgoing byte at the same time as, and asynchronously with, the incoming byte.

Hardware handshaking. Can it be avoided? This is less of a software overhead to manage and most devices these days can handle serial data at a considerable rate.

If all of this seems to be advocating some 'shaky' programming practices, then by all means use the more sophisticated devices (or in the case of the 8051, the inbuilt RS232 communication port).

RS232 involves the use of parallel to serial and serial to parallel converters. In software terms, the accepted technique is:

output	input (when input detected)
load the byte into a register	wait one half bit period
(if required, calculate the parity)	read start bit status
transmit start bit	repeat
repeat	read bit into LSB
wait one bit period	shift all bits left circularly
output the LSB	wait one bit period
shift all bits right circularly	until all data bits received
until all bits sent	store byte into a register
(send parity if required)	(if required, calculate bit period)
(wait one bit period)	(read and compare parity bit)
transmit stop bit	wait one bit period
wait one bit period	read stop bit

Of course, the clever part in this sort of programming is not in the successful, normal situation, but in the handling of errors. What happens if the parity does not match? What happens if the received start and stop bits are not as expected? So much depends on what sort of device is being transmitted to or from the micro-controller that there are many many answers to this sort of problem. Bulletin boards and web sites carry many examples of this type of software and it would be inappropriate merely to copy their sterling efforts in this book.

RS422 Transmission Standard

This serial standard has the virtues of faster data rate (up to 10 Mbits/s) and longer transmission distances (>60 m). It features a balanced double-ended transmission with driver output level in the range of ±2 V to ±6 V. Figure 5.7 shows the nature of the differential output, while Figure 5.8 shows a MAXIM application circuit with the MAX489 receiver/transmitter pair. The highest transmission rate is achieved by having separate pairs of data transmission lines – one set for data in and one set for data out.

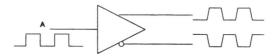

Figure 5.7 RS422 differential output

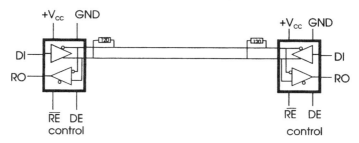

Figure 5.8 MAX489 receiver/transmitter pair

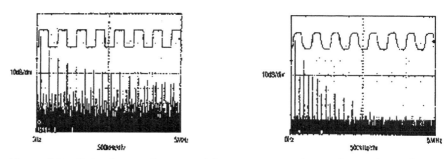

Figure 5.9 Output waveforms and frequency spectrum of non-slew-rate and slew-rate-limited signals

At the highest data rates, a lot of EMI (Electromagnetic Interference) would be generated because of the very fast edges of the square-wave output. To reduce the generated noise, some driver ICs use a *trapezoidal output*. This just means that the edges of the output waveform slope in a controlled way ('slew-rate-limited') to reduce the amount of electrical interference which is produced. Environmentally, and in terms of reliability, this is a good thing, but it does slightly reduce the maximum data rate. Figure 5.9 shows the output waveforms and frequency spectrum of non-slew-rate and slew-rate-limited signals.

RS485 Transmission Standard

This standard is closely related to RS422 except that it is meant for multi-drop applications instead of point-to-point connections. Figure 5.10 shows a typical application.

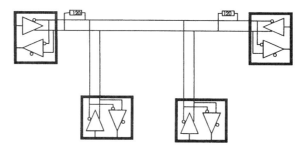

Figure 5.10   A typical RS485 application

I²C Configuration

The I²C or Inter Integrated Circuit bus was originally developed by Philips/ Signetics. It is a simple two-wire interface. The main advantage of the bus is that it is economical with one of the microcontroller's most precious resources – I/O capability. The number of devices which the two lines can connect is limited only by the total line capacitance not exceeding 400 pF. Each of these devices has a unique 7- or 10-bit address and can act as a master or a slave. The full protocol allows for multi-master, multi-slave devices with full collision detection and arbitration for when two masters try to transmit simultaneously. However, many designers keep life simple by using the bus in a single-master (microcontroller) multi-slave configuration.

Transfer of 8-bit data can be accomplished in one of two modes: standard (100 kbits/s) and fast (400 kbits/s).

The protocol offers the following definitions of I²C bus terminology:

Transmitter   The device which sends the data to the bus

Receiver   The device which receives the data from the bus

Master	The device which initiates a transfer, generates clock signals and terminates a transfer
Slave	The device addressed by a master
Multi-master	More than one master can attempt to control the bus at the same time without corrupting a message
Arbitration	A procedure to ensure that if more than one master simultaneously tries to control the bus, only one is allowed to do so and the message is not corrupted

The range of devices with in-built I²C bus is quite comprehensive, and includes: EEPROM, RAM, clock/calendars, LCD display drivers, 4-digit LED display drivers, gate arrays, ADC, DAC, keyboard interfaces, video/audio/radio processors.

Electrical Signals

The devices connect to the bus with open collector (or drain) transistors in the output stage of the integrated circuit and so each of the two wires of the bus must have a pull up resistor in the range of 1–10 kΩ, 4.7 kΩ being typical. Schematically, the bus looks like that shown in Figure 5.11.

Master/Slave Relationships

In Figure 5.11 SDA is the Serial Data line and SCL is the Serial Clock line. SCL is always generated by the masters when they try to initiate a data transfer. The data travels along the bidirectional line SDA. If Master 1 and Master 2 try to move data along SDA at the same time, the first one to output a '0' when the other outputs a '1' loses its access rights and has to wait until the other has finished.

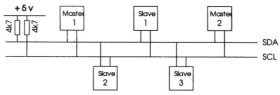

Figure 5.11 I²C bus

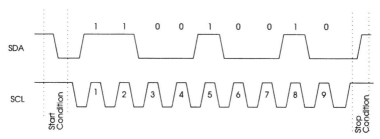

Figure 5.12 Transmission of the binary word 11001001 by I²C bus

At rest (no signal), both SDA and SCL are at logic '1' (+5 V). Normally, SDA can only change when SCL is low – except that

- SCL at '1'; SDA '1' to '0' transition is the START TRANSFER condition

- SCL at '1'; SDA '0' to '1' transition is the STOP TRANSFER condition

Data is transferred in 8-bit bytes, MSB first. This will take 8 clock pulses, and an acknowledge is sent on the ninth clock pulse. Figure 5.12 shows a simple transfer of the binary word 11001001. The acknowledge is obligatory and indicates to the sending device that the receiving device recognizes and accepts the byte. The acknowledging device pulls the SDA line low to acknowledge.

The first byte sent by the master is an address byte. As mentioned, each device has a minimum of a 7-bit address. For each device type, the first four bits are usually fixed, while the lower three are hardwired on the IC to provide a user programmable address. The eighth bit of the transfer is a read/write control. If bit 8 is '0' then the master is WRITING. If bit 8 is '1' then the master is requesting or READING data. Figure 5.13 shows some of the possible combination of transfers.

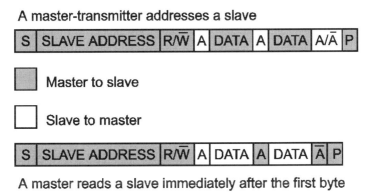

Figure 5.13 Possible combinations of I²C transfers

MCS51 Control Signals

Many of the 8051 family members can transmit and receive I²C signals. It is of course perfectly possible to program a standard member of the family to output bit by bit signals to other I²C devices. This does seem to be the last resort in a microprocessor so well endowed with interface options. Sometimes cost is the main factor and this may be the spur to go down this road. Intel issues an application note (AP476 – available from many of the bulletin boards and web sites) which gives all the code necessary.

Transmission Protocols

The I²C standard defines operation as masters and slaves and the I²C serial I/O can operate in one of four modes:

1. Master Transmitter

2. Master Receiver

3. Slave Transmitter

4. Slave Receiver.

However, in most applications, the MCS51 tends to be used as a bus master controlling the slave peripherals and to simplify the explanation of how to use it, I will only present the data for Modes 1 and 2.

As mentioned, a slave must generate an acknowledge after the reception of each byte and a master must generate an acknowledge after the reception of each byte.

If a receiving device cannot receive the data byte immediately, it can force the transmitter into a wait state by holding SCL Low. If the acknowledge is not received, the master should abort the transfer after an appropriate 'timeout'. There are three I²C Special Function Registers (Table 5.2).

Table 5.2

Name	'552 Symbol	'552 Address
I²C Control	S1CON	D8
I²C Data	S1DAT	DA
I²C Status	S1STA	D9

The Serial Control Register S1CON is used as shown in Table 5.3.

Table 5.3

SCON.7	CR2	Clock set
SCON.6	ENS1	0 = serial I/O disabled: P1.6 and P1.7 are open drain I/O 1 = serial I/O enabled: Output ports must be set to '1'
SCON.5	STA	Start Flag: when set, it will repeatedly generate a START condition
SCON.4	STO	Stop Flag: when set, a STOP condition is generated.
SCON.3	SI	Interrupt Flag: when set an interrupt will be generated if: a start condition is generated or if a data byte has been received or transmitted
SCON.2	AA	Assert Acknowledge: when set an acknowledge is generated if a data byte has been received from a slave
SCON.1	CR1	Clock set
SCON.0	CR0	Clock set: see Table 5.4

Table 5.4

Clock Settings – 12 MHz Crystal				
CR2	CR1	CR0	Bit Frequency	f_{osc} Divided by
0	0	0	47 kHz	256
0	0	1	54 kHz	224
0	1	0	63 kHz	192
0	1	1	75 kHz	160
1	0	0	12.5 kHz	960
1	0	1	100 kHz	120
1	1	0	200 kHz	60
1	1	1	62.5–0.5 kHz	Set by Timer 1 96 × (256–TH1)

The serial status register S1STA only uses the upper 5 bits. These carry a status code which act (if enabled) as an interrupt vector to a service routine. The vectors lie half way between the normal interrupt service vectors, so there is just room in the vector address to fit a jmp location instruction to handle the interrupt. The lower 3 bits are set as 000.

Table 5.5

Master-Transmitter	
08	START transmitted
10	Repeated START transmitted
18	Slave Address and Write transmitted ACK received
20	Slave Address and Write transmitted no ACK received
28	Data transmitted ACK received
30	Data transmitted no ACK received
38	Bus arbitration lost

Table 5.6

Master-Receiver	
38	Arbitration lost while returning ACK
40	Slave Address and Read transmitted ACK received
48	Slave Address and Read transmitted no ACK received
50	Data received ACK returned
58	Data received no ACK returned

S1DAT contains the data to be written out or the data which has been received.

S1ADR can be loaded with the 7-bit slave address to which the controller will respond if programmed as a slave receiver/transmitter. Bit 0 when set determines whether the general call address is to be recognized.

Modes of Operation

The first step must be to set interrupts with instructions such as

```
setb    es1
setb    ea
```

Table 5.7

Setting S1CON (sets a 100 kHz clock rate from a 100 kHz crystal)								
	CR2	ENS1	STA	STO	SI	AA	CR1	CR0
Master Receiver:	1	1	0	0	0	0	0	1

Thereafter, issuing `setb sta` would cause the START sequence to be commenced. Once the I²C master has determined that the bus is free for use and issued the start sequence, the SI flag will be set and the CPU should vector to location 08 as shown in Table 5.7. The code pointed to at this address should load the slave address into the S1DAT register and set the LSB according to whether the data is to be written or read. Lastly, the Interrupt Service Routine should reset SI ready for the next part of the transfer. Once acknowledge has been received from the slave, the SI flag will once again be set and the CPU will vector to either 18, 20 or 38 depending on whether the address was received, not received or lost. These vector addresses must hold code to deal with each situation appropriately (see Table 5.8).

Although this seems complicated, most of the difficulty in programming is in coping with error situations. As always, software gets a bad reputation when code fails to work because the programmer did not cater for every possible outcome.

There are two other tables of data relating to the use of the '552 as a slave. The 8x552 data sheet from Intel or one of the second source suppliers would provide all the register data. If your own assembler is not specifically for the 8x552, then the additional SFR locations and bit names will have to be declared in an equate list.

Capture and Compare Logic

Timer 2 is connected to four 16-bit capture and three 16-bit compare registers. Capture registers save the contents of Timer T2 when a transition (0–1 or 1–0) occurs on its corresponding input pin. These are the alternate functions of Port 4 on a '552 and port 1 on a '751. To continue the analysis on the '751, since individual models differ in detail, essentially there are many common elements between them. If the capture facility is not needed, the four input pins can be used to generate four additional external interrupts. The capture control registers are

CTCON	Capture Control Register
RTE	Reset/Toggle Enable Register
STE	Set Enable Register
TM2IR	Timer 2 Interrupt Flag Register
IP1	Interrupt Priority Register

A repeated external event such as t1, t2, t3 or t4 in Figure 5.14 can easily be measured. A capture register can be used to store the contents of T2 when the 'start' event occurs, and then this number can be subtracted from the updated value of T2 when the 'stop' event occurs.

Table 5.8

Status Vector	I²C Bus Status SIO1 Status	To/From S1DAT	To SIO1			Next Action taken by SIO1 H/W
			STA	STO	SI	
08	Start transmitted	Load Addr+W	X	0	0	Addr+W transmitted; ACK received
10	Repeated Start	Load Addr+W	X	0	0	Addr+W transmitted; ACK received
		or Addr+R	X	0	0	Addr+W transmitted; switch to M/Rx
18	Addr+W transmitted and ACK received	Load data or	0	0	0	Data byte transmitted; ACK received
		no S1DAT action	1	0	0	Repeated start
		no S1DAT action	0	1	0	Stop condition transmitted; STO reset
		no S1DAT action	1	1	0	Stop transmitted, then a Start; STO reset
20	Addr+W transmitted no ACK received	Load data or	0	0	0	Data byte transmitted; ACK received
		no S1DAT action	1	0	0	Repeated Start
			0	1	0	Stop condition transmitted; STO reset
			1	1	0	Stop transmitted, then a Start; STO reset
28	S1DAT data byte transmitted, ACK	Load data or	0	0	0	Data Byte transmitted; ACK received
		no S1DAT action	1	0	0	Repeated Start
		no S1DAT action	0	1	0	Stop condition transmitted; STO reset
		no S1DAT action	1	1	0	Stop transmitted, then a Start; STO reset
30	S1DAT data byte transmitted, no ACK	Load data or	0	0	0	Data Byte transmitted, ACK received
		no S1DAT action	0	0	0	Repeated Start
		no S1DAT action	0	1	0	Stop condition transmitted; STO reset
		no S1DAT action	1	1	0	Stop transmitted, then a Start; STO reset
38	Arbtration lost	no S1DAT action	0	0	0	Start transmitted when the bus frees up
		no S1DAT action	1	0	0	Start transmitted when the bus frees up

Table 5.9 Master Receiver Mode

Status Vector	I²C Bus Status / SIO1 Status	To/From S1DAT	To SIO1				Next Action taken by SIO1 H/W
			STA	STO	SI	AA	
08	Start transmitted	Load Addr+R	X	0	0	X	Addr+R transmitted; ACK received
10	Repeated Start	Load Addr+W	X	0	0	X	Addr+W transmitted; ACK received
		or Addr+R	X	0	0	X	Addr+R transmitted; switch to M/Rx
38	Arbitration lost	no S1DAT action	0	0	0	X	I²C bus released; SIO1 reverts to slavery
		no S1DAT action	1	0	0	X	Start transmitted when the bus frees up
40	Addr+R transmitted and ACK received	no S1DAT action	0	0	0	0	Data byte received; no ACK returned
		no S1DAT action	0	0	0	1	Data byte received; ACK returned
48	Addr+R transmitted no ACK received	no S1DAT action	1	0	0	X	Repeated Start transmitted
		no S1DAT action	0	1	0	X	Stop condition transmitted; STO reset
			1	1	0	X	Stop transmitted, then a Start; STO reset
50	Data byte received, ACK	Read data or	0	0	0	0	Data Byte received; no ACK returned
		Read data	1	0	0	1	Data Byte received; ACK returned
58	Data byte received, no ACK	Read data or	1	0	0	X	Repeated Start transmitted
		Read data or	0	1	0	X	Stop transmitted; STO reset
		Read data	1	1	0	X	Stop tranmitted; then Start; STO reset

Table 5.10 Capture Control Register CTCON (E8)

CTCON.7	CTN3	Capture Register 3 triggered by falling edge on CT3I
CTCON.6	CTP3	Capture Register 3 triggered by rising edge on CT3I
CTCON.5	CTN2	Capture Register 2 triggered by falling edge on CT2I
CTCON.4	CTP2	Capture Register 2 triggered by rising edge on CT2I
CTCON.3	CTN1	Capture Register 1 triggered by falling edge on CT1I
CTCON.2	CTP1	Capture Register 1 triggered by rising edge on CT1I
CTCON.1	CTN0	Capture Register 0 triggered by falling edge on CT0I
CTCON.0	CTP0	Capture Register 0 triggered by rising edge on CT0I

Table 5.11 Reset/Toggle Enable Register RTE (EF)

RTE.7	TP47	If '1' then P4.7 toggles on a match between CM2 and T2
RTE.6	TP46	If '1' then P4.6 toggles on a match between CM2 and T2
RTE.5	RP45	If '1' then P4.5 is reset on a match between CM1 and T2
RTE.4	RP44	If '1' then P4.4 is reset on a match between CM1 and T2
RTE.3	RP43	If '1' then P4.3 is reset on a match between CM1 and T2
RTE.2	RP42	If '1' then P4.2 is reset on a match between CM1 and T2
RTE.1	RP41	If '1' then P4.1 is reset on a match between CM1 and T2
RTE.0	RP40	If '1' then P4.0 is reset on a match between CM1 and T2

Table 5.12 Set Enable Register STE (EE)

STE.7	TG47	Toggle Flip Flops
STE.6	TG47	Toggle Flip Flops
STE.5	SP45	If '1' then P4.5 is set on a match between CM0 and T2
STE.4	SP44	If '1' then P4.4 is set on a match between CM0 and T2
STE.3	SP43	If '1' then P4.3 is set on a match between CM0 and T2
STE.2	SP42	If '1' then P4.2 is set on a match between CM0 and T2
STE.1	SP41	If '1' then P4.1 is set on a match between CM0 and T2
STE.0	SP40	If '1' then P4.0 is set on a match between CM0 and T2

Table 5.13 Interrupt Flag Register TM2IR

TM2IR.7	T2OV	Timer T2 16-bit overflow flag
TM2IR.6	CMI2	CM2 Interrupt Flag
TM2IR.5	CMI1	CM1 Interrupt Flag
TM2IR.4	CMI0	CM0 Interrupt Flag
TM2IR.3	CTI3	CT2 Interrupt Flag
TM2IR.2	CTI2	CT2 Interrupt Flag
TM2IR.1	CTI1	CT2 Interrupt Flag
TM2IR.0	CTI0	CT2 Interrupt Flag

Table 5.14 Timer T2 Interrupt Priority Register IP1 (F8)

IP1.7	PT2	Timer T2 Overflow interrupts priority level
IP1.6	PCM2	Timer T2 Comparator 2 interrupt priority level
IP1.5	PCM1	Timer T2 Comparator 1 interrupt priority level
IP1.4	PCM0	Timer T2 Comparator 0 interrupt priority level
IP1.3	PCT3	Timer T2 Capture register 3 interrupt priority level
IP1.2	PCT2	Timer T2 Capture register 2 interrupt priority level
IP1.1	PCT1	Timer T2 Capture register 1 interrupt priority level
IP1.0	PCT0	Timer T2 Capture register 0 interrupt priority level

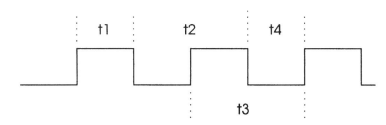

Figure 5.14 Repeated external events

A related function can be achieved with the three 16-bit compare registers. Every time that timer T2 is incremented, the new value is compared with the contents of CM0, CM1 and CM2. If a match is found then the appropriate interrupt flag in TM2IR is set.

Table 5.15

Compare Flag	Effect	Condition IF
CM0	Set Port 4 bits 0–5	corresponding STE bits set
CM1	Reset Port bits 0–5	corresponding RTE bits set
CM2	Toggles Port 4 bits 6, 7	corresponding RTE bits set

I²C Applications

The PCF8593 Low Power Clock Calendar

A simple example can be demonstrated with the PCF8593 Low Power Clock Calendar. Like many VLSI devices, the data sheet is packed with references to the register structure and even the simplest of examples needs a detailed explanation.

This 8-pin SOIC package uses 32.768 kHz crystal as its timing reference and from this derives and maintains a track of:

1. hundreds of a second

2. seconds

3. minutes

4. hours

5. year/date

6. weekdays/months.

Each of these has its own internal register as shown which stores the information in BCD format.

The hundredths, seconds and minutes registers are straightforward two BCD digit registers:

tenths	hundredths	of seconds register 01
tens	units	of seconds register 02
tens	units	of minutes register 03

The hours can be set to work in 12 or 24 hour format. Register 04 is the hours register:

bit 7	mode set	0 = 24 hour	1 = 12 hour
bit 6	am/pm flag	0 = am	1 = pm
bit 5/4	hours tens		
bit 3/0	hours units		

Register 05 handles both days 0–31 and years 0–3

bit 7/6	year
bit 5/4	day tens
bit 3/0	day units

It assumes that year 0 is the leap year and so counts month 2 (February) up to 29.

Register 06 handles weekdays and months:

bit 7/5	weekdays 0–6
bit 4	month tens
bit 3/0	month units

Valid months are from 01 to 12 and it will affect the day count accordingly.

In addition, there is

control register	00
alarm control register	08
alarm registers	09–0F (or free RAM locations)

The IC operates from V_{DD} = 2.5 to 6 V and so is suitable for both of the standard logic supply voltages (3.3 V and 5 V). With a supply current consumption of less than $100 \, \mu A$, it is an ideal candidate for capacitor or battery backup applications. See Figure 5.15.

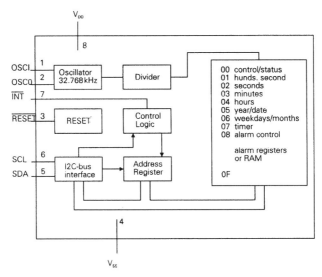

Figure 5.15 The PCF8593 IC

The Control Status register is as shown in Table 5.16.

Table 5.16

D7	Stop counting flag:	0 count pulses; 1 stop counting
D6	Hold last count flag:	0 count; 1 store and hold last count
D5 D4	function mode function mode	00 clock mode 32.768 k 01 clock mode 50 Hz 10 event counter mode 11 test modes
D3	Mask flag:	0 read R05/R06 unmasked; 1 read date and month count directly
D2	Alarm control enable:	0 disabled; 08–0F = RAM 1 enabled; 08 = alarm control register
D1	Alarm flag	50% duty factor minutes flag if alarm enable bit is 0
D0	Timer flag	50% duty factor seconds flag if alarm enable bit is 0

The Alarm Control Register in Clock mode is an 8-bit register at location 08 (Table 5.17).

Table 5.17

D7	Alarm interrupt enable	0 = alarm flag no interrupt 1 = alarm flag interrupt
D6	Timer Alarm Enable	0 = no timer alarm 1 = timer alarm
D5 D4	Clock alarm function Clock alarm function	00 no clock alarm 01 daily alarm 10 weekday alarm 11 dated alarm
D3	Timer Interrupt Enable	0 = timer flag no interrupt 1 = timer flag interrupt
D2 D1 D0	Timer function Timer function Timer function	000 no timer 001 hundreds of a second 010 seconds 011 minutes 100 hours 101 days 110 not used 111 test mode: all counter in parallel

An alarm signal is generated when the contents of the alarm registers matches exactly the contents of the involved counter registers. The year and weekday bits are ignored in a dated alarm. A daily alarm ignores the month and date bits. When a weekday alarm is selected the contents of the alarm weekday/month register (register 0E) will select the weekdays on which an alarm is activated:

bit:		7	6	5	4	3	2	1	0
weekday enabled:		n/a	6	5	4	3	2	1	0

Programming

The PCF8593 has a fixed I^2C address of 1010001x. The last bit determines whether the master is writing (0) or reading (1). Hence the addresses to be used by the master are $C2_H$ and $C3_H$.

The micro can either:

● write to the PCF8593 to initialize or set registers or set alarms

or

● read from the PCF8593 to determine time or date information in the following communication sequences:

S = Start

A = Acknowledge

P = Stop

1 = No Acknowledge

To set the PCF8593 into timer mode:

S	C2	A	00	A	08	A	P

C2 addresses the PCF8593 in write mode to address 00, sending data 08.

To set the time and date as 10:23:06.10 on Tuesday January 30 of a leap year:

S	C2	A	01	A	10	A	06	A
23	A	10	A	30	A	41	A	P

To read the hours and minutes:

S	C2	A	03	A	S	C3	A	mm
A	hh	1	P					

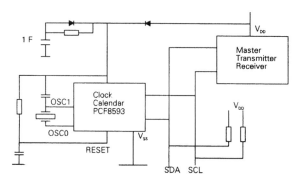

Figure 5.16   Circuit for the PCF8593

Note that under the I²C protocol, the **A** following the mm transmission from the slave has to be provided by the master. All the other Acknowledge **A**s are sent from the Slave PCF 8593. An application circuit is shown in Figure 5.16.

MCS51 (80552) Software Chunks

As previously discussed, the 80552 with an I²C port would first need to have the appropriate ports set to logic 1.

```
setb      P1.6      ; SCL
setb      P1.7      ; SDA
```

It is then necessary to enable interrupts

```
setb      EA
setb      ES1
```

and to start a transmission

```
setb      STA
```

When the I²C bus is clear, the CPU will vector to program memory 08 where there would usually be a `jmp` to a less congested part of program memory. This would normally be down with the subroutines at the end of the program. The '08' Interrupt Service Routine has to write a 'C2' into the SDAT register to generate a Start condition.

```
mov      S1DAT, #C2H
clr      SI                ; reset the Interrupt Flag
ret
```

If an ACKnowledge is received, all is well and the act of receiving an ACK will cause a vector to location 18_H. The code at 18 will cause a jump into the Interrupt Service Routine proper. This Service routine will be called twice in our trivial example; once to send 00 and once to send 08. Thus the code is now getting more complicated, since we must devise a method of 'remembering' which byte we have transferred. The usual technique is to use a register or a memory location to act as a pointer to a table of data or as an offset within that table. Suppose we labelled a suitable location as setup_marker to hold a down counter:

```
mov  R1,   #setup_marker
mov  @R1,  #2                    ; 2 digits to download
                                 ; stored in 'setup_marker'
```

and a table in memory could be created with the simple define byte code:

```
setup_table:    db 08, 00
```

In this example, the data goes in reverse order so the Interrupt Service Routine would look something like:

```
push  r1
mov   r1, #setup_marker
cjne  @r1, 0, done        ; if all bytes sent
mov   dptr, #setup_table
mov   a, @r1
dec   @r1                 ; downcount
mov   a, @a + dptr        ; fetch the next byte
mov   S1DAT, a
clr   SI                  ; done this byte
pop   r1                  ; restore Register R1
reti

done:

pop   r1
clr   SI
setb  STO                 ; STOp the transmission
reti
```

As already mentioned, the difficulties come when an error condition occurs.

Receiving data is very similar, except that the '552 has to change into a Master Receiver half way through the transmission.

U4224B

Of course, if measurement or display of time is important, then a more interesting solution is to decode time signals from the many low-frequency transmitters which operate around the world. One such IC is the U4224B from Telefunken Semiconductors. This IC is capable of decoding signals from

DCF77	Mainflingen, Germany	77.5 kHz	50 kW
MSF	Teddington, England	60 kHz	50 kW
WWVB	Fort Collins, USA	60 kHz	10 kW
JG2AS	Sanwa, Japan	40 kHz	10 kW

Each shares the same characteristic that the transmission sequence has an index count of 1 second, and during the 1 minute time frame, transmits minute, hour, day of week, day of month, year and:

Summertime flag	(D)
BST flag	(UK)
Daylight saving information	(USA)
Leap year flag	(USA)

The application circuit is quite minimal, consisting as it does of the IC, two crystals, four capacitors and a ferrite aerial. Interface to the microprocessor is by three lines, and of these, two are for U4224B control and one is the Time Code Output signal. TCO is a digital series of 0s and 1s which are distinguished by the pulse width.

The USA and Japan additionally have six 200 ms pulse width position markers every 10 seconds. The German start-of-minute frame is recognized by a missing pulse, whereas all of the others have a unique pulse width.

This section was just included for completeness for those who are interested in national transmitted standards. Decoding hardware is cheap and easy to implement.

Table 5.18

Country	0	1	Frame Reference
Germany	100 ms	200 ms	200 ms
UK	100 ms	200 ms	500 ms
USA	800 ms	500 ms	200 ms
Japan	800 ms	500 ms	200 ms

Decoding software would just rely on being able to distinguish between pulse widths and to be able to count. The data is output in BCD format and is found at the bit positions shown in Table 5.19.

Table 5.19

	year	month	day	weekday	hour	minute	
Germany	50–58	45–49	36–41	42–44	29–34	21–27	LSB first
UK	17–24	25–29	30–35	36–38	39–44	45–52	MSB first
USA	45–54			22–33	12–18	1–8	MSB first
Japan				22–33	12–18	1–8	MSB first

Figure 5.17 shows the application.

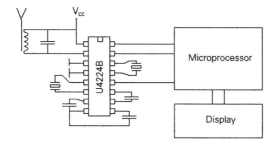

Figure 5.17 Circuit for the U4224B

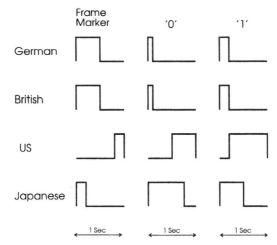

Figure 5.18 Pulse systems used by time signal transmitters

Chapter 6

LCD Text Display Panels

Many microcontroller projects need some sort of output display unit. A simple and cheap option is based on the Hitachi 44780 IC. This drives an LCD dot matrix panel which is capable of alphanumeric and other characters. The commonly available formats are:

Characters	8	16	16	20	24	40	40
Lines	1	1	2	1	1	2	4

Depending on the type of LCD, the module (controller IC plus LCD display) will require single or dual supplies. Interfacing is very easy because of the common layout.

Table 6.1

Pin	Function	
1	V_{SS}	GND Supply Voltage
2	V_{DD}	+5 Supply Voltage
3	V_O	LCD contrast control
4	RS	Register Select: 0 = Instruction Register 1 = Data Register
5	R/W	Read/Write: 0 = Write 1 = Read
6	E	Enable: Data is written at the falling edge of E
7–14	DB0-DB7	Data in

If a backlight is fitted, it is usually supplied or controlled by additional pins 15 and 16.

Because I/O pins are a scarce resource in most simple microcontrollers, it is a useful feature of the module that it can be driven via an 8-bit data bus (DB0–DB7) or via

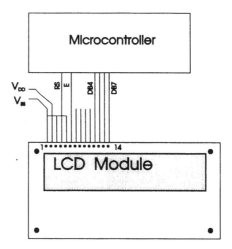

Figure 6.1 Interface for LCD

a 4-bit data bus (DB4–DB7). The software examples below bring the total interface down to 6 signal pins: RS, R/W, DB4, DB5, DB6 and DB7.

Most users wish to write to the module. It is not such a common need to read back what is already there. This simplifies the interface to that of Figure 6.1. The disadvantage is that the user cannot read whether the module is still busy dealing with the last character sent. In practice, this is not a problem since there is no point updating the display faster than it can be read, so simple delays (c. 100 µs) can be used between each character sent.

At power on, the module will reset itself. Thereafter, the microcontroller must configure the module to the correct format. Once set up for 4-bit transfer, the data and control bytes must be sent upper-nibble then lower-nibble to input pins DB4-DB7. The correct procedure is:

1. Power on.

2. Wait for more than 15 ms.

The 16 × 2 module (code on page 164) was set up with the code sequence:

```
02      Set Data length = 4 bits
28      4 bits, 2 lines, 5 × 7 font
0E      Display On, Cursor On
06      Cursor > , address >
```

Thereafter, the data character set is essentially ASCII.

Table 6.2

Instruction:	RS	D7	D6	D5	D4	D3	D2	D1	D0
Clear Display	0	0	0	0	0	0	0	0	1
Return Home	0	0	0	0	0	0	0	1	0
Set Entry Mode	0	0	0	0	0	0	1	I/D	S
Display Control	0	0	0	0	0	1	D	C	B
Cursor & Shift Ctrl	0	0	0	0	1	S/C	R/L	*	*
Function Set	0	0	0	1	DL	N	F	*	*
Set CG RAM addr	0	0	1	<	Address				>
Set DD RAM addr	0	1	<	Address					>
Write Data	1	<	Data						>

Table 6.3

I/D	Cursor Move Direction	0 = Increment	1 = Decrement
S	Display Shift	0 = Still	1 = Accompanies Shift
D	Display On/Off	0 = Off	1 = On
C	Cursor On/Off	0 = Off	1 = On
B	Cursor Blink	0 = No Blink	1 = Blink
S/C	Move Cursor & Shift Display	0 = Cursor Move	1 = Display Shift
R/L	Shift Direction	0 = Left Shift	1 = Right Shift
DL	Data Length	0 = 4 bits	1 = 8 bits
N	Number of Lines	0 = 1 line	1 = 2 lines
F	Character Font	0 = 5 × 7 dots	1 = 5 × 10 dots
CG = Character Generator RAM		DD = Display Data RAM	

In a 16×2, 20×2 or 40×2 display, the DD RAM address must be set at 40 to write to the second line. A 20×4 LCD starts each line at addresses.

- 1   00

- 2   40

- 3   14

- 4   54

Of course, the interfacing becomes even more minimal if an I²C interface is used. In many instances, this causes the cost to go up. Cost is largely a factor of units sold. At present, the parallel 4/8-bit module is the cheapest variant. If I²C is needed, it might perhaps be cheaper to have one microcontroller (83C751 or PIC16C54) interfaced to the module, itself providing an I²C interface to the rest of the world. The microcontroller would then have to mimic a slave I²C device.

MCS51 Code

```
; LCD16 × 2 mrj '94
; A demonstration programme for the Hitachi
; 16 char × 2 line LCD dot matrix display

Enable          reg P1.5
Data_Mode       reg P1.4
Button          reg P3.2

start: .equ 0h
    .org  start
    jmp   main
    .org  start + 20h ; clear the interrupt region

main:
    call  initialise_lcd
    call  clear_display

    mov   dptr, #message_1
    call  display_data
    call  clear_display

    mov   dptr, #message_2
    call  display_data
    call  move_to_second_line

    mov   dptr, #message_3
    call  display_data
    jmp $

; ***** SUBROUTINES ******
    ;uses A, R7
delay:   mov    r7, #255
dly01:   djnz   r7, dly01
    ret
```

```
display_data:
    ; displays message pointed to in program memory
    ; by DPTR
    ; uses DPTR, A, R7
disp01:   mov a, #0
    movc  a, @a + dptr
    cjne  a, #00, disp02
    ret
disp02:   call char_to_lcd
    inc   dptr
    jmp   disp01

send_codes:
    ; sends a sequence of codes pointed to in
    ; program memory by the DPTR
    ; uses DPTR, A, R7
    clr   Data_Mode
sc01:     mov a, #0
    movc  a, @a + dptr
    cjne  a, #00, sc02
    setb  Data_Mode
    ret
sc02:     call char_to_lcd
    inc   dptr
    jmp   sc01

char_to_lcd:
    ; Writes a byte of data to the LCD
    ; in two nibbles- uses DPTR, A, R7
    mov   a, #0
    movc  a, @a + dptr
    anl   a, #f0h
    swap  a
    anl   p1, #f0h; keep upper nibble the same
    orl   p1, a
    call  enable_sequence
    mov   a, #0
    movc  a, @a + dptr
    anl   a, #0fh
    anl   p1, #f0h
    orl   p1, a
    call  enable_sequence
    ret

enable_sequence:
    ; puts the Enable pin through its correct
```

```
    ; sequence
    ; uses A, R7
    call  delay
    clr   Enable
    call  delay
    setb  Enable
    call  delay
    ret

move_to_first_line:
    ; moves cursor to the head of the
    ; first line and clears it - uses DTPR, A, R7
    mov   dptr, #first_line
    call  send_codes
    ret

move_to_second_line:
    ; moves cursor to the head of the
    ; second line and clears it - uses DPTR, A, R7
    mov   dptr, #second_line
    call  send_codes
    ret

clear_display:
    ; uses DPTR, A, R7
    mov   dptr, #cls_code
    call  send_codes
    ret

initialise_lcd:
    ; passes through the initialisation
    ; sequence of the LCD - uses DPTR, A, R7
    mov   p1, #0
    setb  Enable
    clr   Data_Mode
    orl   p1, #00000010b; data length = 4 bits
    call  enable_sequence
    call  delay
    call  delay
    mov   dptr, #lcd_init_codes
    call  send_codes
    ret

lcd_init_codes:
    db 28h      ; 4 bits, 2 lines, 5 × 7 font
    db 0eh      ; display on, cursor on
```

```
        db 06h      ; cursor →, address →
        db 00

    first_line:
        db 02h      ; code for first line
        db 00

    second_line:
        db c0h      ; code for the second line
        db 00

    cls_code:
        db 01       ; clear and return home
        db 00

message_1:
    db 'Mike James', 00

message_2:
    db '*** MCS 51 ***', 00

message_3:
    db '**** PIC ****', 00

.end
```

PIC Code

```
; LCD.TXT MRJ 94
; This code is suitable for the HITACHI
; (and compatible) range of Alphanumeric
; LCD display panels.
; LCD.INI contains TWO preset constants -
; RS = 5 : E = 4
; and uses TWO registers : 1E & 1F
; other than that it contains the LCD
; initialisation codes
; LCD.MAC contains 3 accessible MACROS :
; lcd_control_mode permits control code entry
; to the LCD
; lcd_data_mode permits data entry into the LCD
; lcd_write 'x' will write the character in quotes
; LCD.SUB contains 3 accessible SUBROUTINES :
; row1 clear LCD screen and home the cursor
```

```
; row2 move the cursor to the head of row 2
; char_to_lcd write the character in W to the LCD.

; LCD.INI MRJ 11/94

RS          equ     5
Enable      equ     4
CHAR        equ     h'1e'
DLYreg      equ     h'1f'

bsf         PortB, Enable        ; enable line is
                                 ; normally high
   lcd_control_mode
   movlw       2                 ; data transfer = 2 × 4
                                 ; bit nibbles
   iorwf       PortB, f
   call lcd_enable_sequence
   lcd_small_delay
   lcd_small_delay               ; let things settle
   lcd_write H'28'               ; 2 lines; 4 bits; 5 × 7
                                 ; font
   lcd_write   H'0E'             ; display on ; cursor on
   lcd_write   H'06'             ; cursor > ; address >
   lcd_data_mode

; ***** MACROS ******

   lcd_small_delay MACRO         ; 256 iterations
   local       lcd_sd
   movlw       H'ff'
   movwf       DLYreg
   lcd_sd      decfsz      DLYreg, f
   goto lcd_sd
   ENDM

lcd_data_mode MACRO              ; change to data entry
   bsf   PortB, RS
   ENDM

lcd_control_mode MACRO           ; change to control
                                 ; entry
   bcf   PortB, RS
   ENDM

lcd_write       MACRO arg1       ; more expressive write
command
```

```
        movlw        arg1
        call  char_to_lcd
        ENDM

; WDT_LCD.ASM drives an LCD under the SLEEP command
; MRJ 19/11/94
; requires an LCD connected to PortB :
; RB0    D0/4   PIN
; RB1    D1/5   PIN
; RB2    D2/6   PIN
; RB3    D3/7   PIN
; RB4    E      PIN
; RB5    RS     PIN

bsf       STATUS, RP0           ; page 1
movlw     B'00000000'
movwf     TrisB                 ; all PortB = output
movlw     B'00001110'           ; WDT on @ 1/64 rate
movwf     OPTI
bcf       STATUS, RP0           ; page 0

loop      sleep
    call  message1
    sleep
    call  message2
    sleep
    call  message3

    goto  loop

message1
    call  row1
    lcd_write ' '
    lcd_write ' '
    lcd_write 'M'
    lcd_write 'I'
    lcd_write 'k'
    lcd_write 'e'
    lcd_write ' '
    lcd_write ' '
    call row2
    lcd_write ' '
    lcd_write 'J'
    lcd_write 'a'
    lcd_write 'm'
    lcd_write 'e'
```

```
            lcd_write 's'
            lcd_write ' '
            lcd_write ' '
            return

    message2
        call row2
            lcd_write ' '
            lcd_write ' '
            lcd_write 'P'
            lcd_write 'I'
            lcd_write 'C'
            lcd_write ' '
            lcd_write ' '
            lcd_write ' '
        call row2
            lcd_write ' '
            lcd_write ' '
            lcd_write '8'
            lcd_write '0'
            lcd_write '5'
            lcd_write '1'
            lcd_write ' '
            lcd_write ' '
            return

    message3
        call row1
            lcd_write 'L'
            lcd_write 'C'
            lcd_write 'D'
            lcd_write ' '
            lcd_write ' '
            lcd_write ' '
            lcd_write ' '
            lcd_write ' '
        call row2
            lcd_write 'd'
            lcd_write 'i'
            lcd_write 's'
            lcd_write 'p'
            lcd_write 'l'
            lcd_write 'a'
            lcd_write 'y'
            lcd_write ' '
            return
```

```
; LCD.SUB MRJ 11/94

row1      lcd_control_mode      ; change to LCD row 1
    lcd_small_delay
    lcd_write   H'01'
    lcd_data_mode
    lcd_small_delay
    return

row2      lcd_control_mode      ; change to LCD row 2
    lcd_small_delay
    lcd_write H'C0'
    lcd_data_mode
    lcd_small_delay
    return

char_to_lcd      ; write character in W register
    movwf         CHAR
    movlw         H'f0'
    andwf         PortB, f      ; clear lower nibble
                                ; PortB
    swapf         CHAR, w       ; collect upper data
                                ; nibble
    andlw         H'0f'
    iorwf         PortB, f      ; send it
    call lcd_enable_sequence ; latch it into LCD
    movlw         H'f0'
    andwf         PortB, f      ; reclear lower nibble
    movf          CHAR, w       ; collect lower nibble
    andlw         H'0f'
    iorwf         PortB, f      ; send it
    call lcd_enable_sequence ; latch it into
                                ; LCD
    return

lcd_enable_sequence              ; latch data into LCD
    lcd_small_delay
    bcf   PortB, Enable          ; enable bit down
    lcd_small_delay
    bsf   PortB, Enable          ; enable bit back up
    lcd_small_delay
    return

    end
```

Appendices

A MCS51 Instruction Set

Notes

Rn	Register R7–R0 of currently selected register bank
direct	8-bit internal data location's address
@Ri	Internal RAM location 0–255 addressed indirectly through R7–0
#data	8-bit constant included in instruction
#data16	16-bit constant included in instruction
addr16	16-bit destination address (used in LCALL & LJMP)
addr11	11-bit destination address (used in ACALL & AJMP)
rel	Signed (2's complement) 8-bit offset byte (SJMP)
bit	Direct addressed bit in internal data RAM or Special Function Register

Arithmetic Operations

ADD	A,Rn	Add register n to A
ADD	A,direct	Add direct byte to A
ADD	A,@Ri	Add indirect RAM to A
ADD	A,#data	Add immediate data to A
ADDC	A,Rn	Add with carry Register n to A
ADDC	A,direct	Add with carry direct byte to A
ADDC	A,@Ri	Add with carry indirect RAM to A
ADDC	A,#data	Add with carry immediate data to A
SUBB	A,Rn	Subtract with borrow Register n from A
SUBB	A,direct	Subtract with borrow direct byte from A
SUBB	A,@Ri	Subtract with borrow indirect RAM from A

SUBB	A,#data	Sub. with borrow immediate data from A
INC	A	Increment Accumulator
INC	Rn	Increment Register n
INC	direct	Increment direct byte
INC	@Ri	Increment indirect RAM
DEC	A	Decrement Accumulator
DEC	Rn	Decrement Register n
DEC	direct	Decrement direct byte
DEC	@Ri	Decrement indirect RAM
INC	DPTR	Increment Data Pointer
MUL	AB	Multiply A & B
DIV	AB	Divide A by B
DA	A	Decimal Adjust Accumulator

Logical Operations

ANL	A.Rn	AND Register n to A
ANL	A,direct	AND direct byte to A
ANL	A,@Ri	AND indirect RAM to A
ANL	A,#data	AND immediate data to A
ANL	direct,A	AND A to direct byte
ANL	direct,#data	AND immediate data to direct byte
ORL	A.Rn	OR Register n to A
ORL	A,direct	OR direct byte to A
ORL	A,@Ri	OR indirect RAM to A
ORL	A,#data	OR immediate data to A
ORL	direct,A	OR A to direct byte
ORL	direct,#data	OR immediate data to direct byte
XRL	A.Rn	Ex-OR Register n to A
XRL	A,direct	Ex-OR direct byte to A
XRL	A,@Ri	Ex-OR indirect RAM to A
XRL	A,#data	Ex-OR immediate data to A

XRL	direct,A	Ex-OR A to direct byte
XRL	direct,#data	Ex-OR imm. data to direct byte
CLR	A	Clear A
CPL	A	Complement A
RL	A	Rotate A Left
RLC	A	Rotate A Left through Carry
RR	A	Rotate A Right
RLC	A	Rotate A Right through Carry
SWAP	A	Swap nibbles within A

Data Transfer

MOV	A,Rn	Move Register n to A
MOV	A,direct	Move direct byte to A
MOV	A,@Ri	Move indirect RAM to A
MOV	A,#data	Move immediate data to A
MOV	Rn,A	Move A to Register n
MOV	Rn,direct	Move direct byte to Register n
MOV	Rn,#data	Move immediate data to Register n
MOV	direct,A	Move A to direct byte
MOV	direct,Rn	Move Register n to direct byte
MOV	direct,direct	Move direct byte to direct byte
MOV	direct,@Ri	Move indirect RAM to direct byte
MOV	direct,#data	Move immediate data to direct byte
MOV	@Ri,A	Move A to indirect RAM
MOV	@Ri,direct	Move direct byte to indirect RAM
MOV	@Ri,#data	Move immediate data to indirect RAM
MOV	DPTR,#data16	Load data ptr with a 16-bit const
MOVC	A,@A+DPTR	Move code byte rel. to DPTR to A
MOVC	A,@A+PC	Move code byte relative to PC to A
MOVX	A,@Ri	Move external RAM (8-bit) to A
MOVX	A,@DPTR	Move external RAM (16-bit) to A
MOVX	@Ri,A	Move A to external RAM (8-bit)

MOVX	@DPTR,A	Move A to external RAM (16-bit)
PUSH	direct	Push direct byte onto the stack
POP	direct	Pop direct byte from stack
XCH	A,Rn	Exchange register n with A
XCH	A,direct	Exchange direct byte with A
XCH	A,@Ri	Exchange indirect RAM with A
XCHD	A,@Ri	Exch. low-order digit ind. RAM with A

Boolean Variable Manipulation

CLR	C	Clear carry
CLR	bit	Clear direct bit
SETB	C	Set carry
SETB	bit	Set direct bit
CPL	C	Complement carry
CPL	bit	Complement direct bit
ANL	C,bit	AND direct bit to carry
ANL	C,/bit	AND complement of direct bit
ORL	C,bit	OR direct bit to carry
ORL	C,/bit	OR complement of direct bit
MOV	C,bit	Move direct bit to carry
MOV	bit,C	Move carry to direct bit
JC	rel	Jump if carry set
JNC	rel	Jump if carry not set
JB	bit,rel	Jump if direct bit set
JNB	bit,rel	Jump if direct bit not set
JBC	bit,rel	Jump if direct bit is set and clear bit

Program Branching

ACALL	addr11	Absolute subroutine call
LCALL	addr16	Long subroutine call
RET		Return from subroutine
RETI		Return from interrupt
AJMP	addr11	Absolute jump

LJMP	addr16	Long jump
SJMP	rel	Short jump (relative addr)
JMP	@A+DPTR	Jump indirect relative to the DPTR
JZ	rel	Jump if A is zero
JNZ	rel	Jump if A is not zero
CJNE	A,direct,rel	Comp dir byte to A and jump if not eq.
CJNE	A,#data,rel	Comp imm to A and Jump if not eq.
CJNE	Rn,#data,rel	Comp imm to reg n and jump if not eq.
CJNE	@Ri,#data,rel	Comp imm to ind and jump if not eq.
DJNZ	Rn,rel	Dec register n and jump if not zero
DJNZ	direct,rel	Dec direct byte and jump if not zero
NOP		No operation

The Special Function Register

Table A.1 The Special Function Register

Address	Symbol	Address	Symbol
F8H		98H	SCON
F0H	B register	90H	P1
E8H		8DH	TH1
E0H	ACCumulator	8CH	TH0
D8H		8BH	TL1
D0H	PSW	8AH	TL0
C8H		89H	TMOD
C0H		88H	TCON
B8H	IP	83H	DPH
B0H	P3	82H	DPL
A8H	IE	81H	SP
A0H	P2	80H	P0
99H	SBUF		

B register This is mostly used in multiply and divide operations. If not required for these functions, the memory space is available.

PSW (Program Status Word)   CY AC F0 RS1 RS0 0V – P

Table A.2

CY	PSW.7	Carry Flag: set/cleared by hardware or software during certain arithmetic and logical instructions
AC	PSW.6	Auxiliary Carry Flag: set/cleared by hardware during addition or subtraction instructions to indicate carry or borrow out of bit 3
F0	PSW.5	Flag0: set/cleared/tested by software as a user-defined status flag
RS1	PSW.4	Register bank select control bits
RS0	PSW.3	
0V	PSW.2	Overflow flag: set/cleared by hardware during arithmetic instructions to indicate overflow conditions
–	PSW.1	(reserved)
P	PSW.0	Parity Flag: set/cleared by hardware each instruction cycle to indicate an odd/even number of 'one' bits in the accumulator

IP (Interrupt Priority)   These set the priority of the 5 interrupt sources. There are two priority levels: low and high. A 0 in the bit position is a low priority, while a 1 sets a high priority. Within each priority level, the order of evaluation of simultaneous interrupts is

IE0	IP.0	External interrupt 0
TF0	IP.1	Timer 0 interrupt
IE1	IP.2	External interrupt 1
TF1	IP.3	Timer 1 interrupt
RI + TI	IP.4	Serial port interrupt

P3 (Port 3) A general-purpose Input/Output port. It has the following alternate functions shown in Table A.3.

Table A.3

RD	P3.7	Read data control output. Active low pulse generated by hardware when external data memory is read
WR	P3.6	Write data control output. Active low pulse generated by hardware when external data memory is writtten
T1	P3.5	Timer/Counter 1 external input or test pin
T0	P3.4	Timer/Counter 0 external input or test pin
INT1	P3.3	Interrupt 1 input pin. Low-level or falling edge trigger
INT0	P3.2	Interrupt 0 input pin. Low-level or falling edge trigger
TXD	P3.1	Transmit Data pin for serial port in UART mode. Clock output in shift register mode
RXD	P3.0	Receive Data pin for serial port in UART mode. Data I/O pin in shift register mode

IE Interrupt Enable register

Table A.4

EA	IE.7	Enable All control bit: cleared by software to disable all interrupts, independent of the state of IE.4–IE.0
–	IE.6	
–	IE.5	
ES	IE.4	Enable serial port control bit: set/cleared by software to enable/disable interrupts from TI or RI flags
ET1	IE.3	Enable Timer 1 control bit: set/cleared by software to enable/disable interrupts from Timer/Counter 1
EX1	IE.2	Enable External interrupt 1 control bit: set/cleared by software to enable/disable interrups from INT1
ET0	IE.1	Enable Timer 0 control bit: set/cleared by software to enable/disable interrupts from Timer/Counter 0
EX0	IE.0	Enable External interrupt 0 control bit: set/cleared by software to enable/disable interrupts from INT0

P2   Port 2   SCON   Serial Port Control/Status Register.

Table A.5

SM0	SCON.7	Serial port mode control bit 0: set/cleared by software
SM1	SCON.6	Serial port mode control bit 1: set/cleared by software
SM2	SCON.5	Serial port mode control bit 2: set by software to disable reception of frames for which bit 8 is zero
REN	SCON.4	Receiver Enable control bit: set/cleared by software to enable/disable serial data reception
TB8	SCON.3	Transmit bit 8: set/cleared by hardware to determine state of ninth data bit transmitted in 9-bit UART mode
RB8	SCON.2	Receive bit 6: set/cleared by hardware to indicate state of ninth data bit received
TI	SCON.1	Transmit Interrupt flag: set by hardware when byte transmitted. Cleared by software after servicing
RI	SCON.0	Receive interrupt flag: set by hardware when byte received. Cleared by software after servicing

SM0 SM1
0   0 Shift register I/O expansion
0   1 8-bit UART variable data rate
1   0 9-bit UART fixed data rate
1   1 9-bit UART variable data rate

P1 (State of Port 1)   TCON   Timer/Counter Control/Status Register

Table A.6

TF1	TCON.7	Timer 1 overflow flag: set by hardware on Timer/Counter overflow. Cleared when interrupt processed
TR1	TCON.6	Timer 1 run control bit: set/cleared by software to turn Timer/Counter on/off
TF0	TCON.5	Timer 0 overflow flag: set by hardware on Timer/Counter overflow. Cleared when interrupt processed
TR0	TCON.4	Timer 0 run control bit: set/cleared by software to turn Timer/Counter on/off
IE1	TCON.3	Interrupt 1 edge flag: set by hardware when external interrupt edge detected. Cleared when interrupt processed
IT1	TCON.2	Interrupt 1 type control bit: set/cleared by software to specify falling edge/low level triggered external interrupts
IE0	TCON.1	Interrupt 0 edge flag: set by hardware when external interrupt edge detected. Cleared when interrupt processed
IT0	TCON.0	Interrupt 0 type control bit: set/cleared by software to specify falling edge/low level triggered external interrupts

SP (Stack Pointer)   At a reset, the stack pointer is set to 07. When data is pushed onto the stack, the stack is first incremented, then data is stored at that stack pointer address. A pop retrieves data from the current SP address and then decrements the stack pointer. Note that the default setting for the SP will cause it to overwrite the Register bank 1.

P0 (Port 0)

B   Number Systems

Computers deal in binary quantities, i.e. a 'low' voltage (0 V–0.8 V) is a logic 'zero' while a 'high' voltage (3.4 V–5.0 V) is a logic 'one'. Humans work in a decimal system, and as such you probably learned your maths as Hundreds, Tens and Units (HTU). Mathematically, we could express this as 10^2, 10^1, 10^0, so a sum would be:

	H	T	U
	10^2	10^1	10^0
	4	6	2
+	1	8	3
	6	4	5

Note that adding 6 tens to 8 tens generates a 'carry', i.e. it overflows the capability of the tens column.

The same happens in the binary scale. We do not have Hundreds, Tens and Units, but we do have powers of 2:

8	4	2	1
2^3	2^2	2^1	2^0

A decimal system has 10 symbols (0123456789) while the binary system has 2 (01).

Conversion: Binary → Decimal

A binary number can simply be converted to decimal by noting the position and value of the 1s:

2^7	2^6	2^5	2^4	2^3	2^2	2^1	2^0	
128	64	32	16	8	4	2	1	
0	1	1	0	0	1	0	1	$= 64 + 32 + 4 + 1 = 101_{10}$
0	1	1	1	1	0	0	0	$= 64 + 32 + 16 + 8 = 120_{10}$
0	0	0	1	0	1	0	0	$= 16 + 4 = 20_{10}$

Conversion: Decimal → Binary

A common way of converting decimal numbers to binary is to use the 'repeated division by 2 method'. For example, to convert 207_{10} into binary:

```
2)    207
2)    103    r1
2)     51    r1
2)     25    r1
2)     12    r1
2)      6    r0
2)      3    r0
        1    r1
```

The initial number is repeatedly divided by 2 until either a '1' or a '0' is left. At each stage, the remainder is written down, i.e. 207 divided by 2 is 103 r 1 etc. The answer is then read from *bottom to top*, i.e. $207_{10} = 11001111_2$.

Questions

1. Convert the following binary numbers to decimal:

   (a) 10110   (d) 1000101

   (b) 11000011   (e) 11010111

   (c) 10100101   (f) 01111111

2. Convert the following decimal numbers to binary:

   (a) 112   (b) 193   (c) 69

   (d) 87   (e) 29   (f) 237

Hexadecimal Notation

Computers work in binary quantities, and also handle words of various sizes. These could be 4, 8, 16, 32 or even 64 bits wide. In this book, we are concerned mostly with 8-bit processors. 8 bits are known as a byte, while 4 bits are known as a nibble.

4 bits contain 16 possible binary numbers, so it has become convenient to use a numbering system called *hexadecimal*. In this system there are 16 single symbols: 0 1 2 3 4 5 6 7 8 9 A B C D E F. Our 'HTU' sum has now become

16^3	16^2	16^1	16^0
4096	256	16	1

Conversion: Decimal → Hexadecimal

Given a number such as 237, it will be found that 237 divided by 16 is 14 r 13. 14 has a hexadecimal value of E. 13 has a hexadecimal value of D. So, 237_{10} = ED_{16}.

Conversion: Hexadecimal → Decimal

Conversion of a number such as 7A requires the use of both number systems, i.e. A has a decimal value of 10, 7 in the 16_1 column has a decimal value of $7 \times 16 = 112$. $7A_{16} = 112_{10} + 10_{10} = 122_{10}$.

Of course, the simple way to convert is to use the convert facility available on most calculators. But there is no reason for not knowing the mechanism behind the conversion principle.

This section is concerned with the manipulation of 8-bit binary and hexadecimal numbers. It deals with the various ways in which numbers can be combined.

Questions

1. convert into hexadecimal:

   (a) 63   (b) 99   (c) 180   (d) 200

2. convert into decimal:

   (a) 22   (b) C9   (c) 1B   (d) F3

Logic

There are four main logical fuctions: AND, OR, NOT, EXCLUSIVE OR. They are defined as follows:

AND   The output is true if and only if all inputs are true:

    X = A . B   . is the AND function

OR   The output is true if any of the inputs are true.

    X = A + B   + is the OR function

NOT   The output is true if the output is false.

$X = \overline{A}$ $\overline{}$ is the NOT function

EXCLUSIVE OR   The output is true if only one of the inputs is true.

$X = A \oplus B$ $\oplus$ is the EX-OR function

The following examples show the effects of two 8-bit numbers being combined.

AND

00111010

<u>11110110</u>

00110010

Only where both digits are a 1 will the output be a 1.

OR

11001100

<u>10101010</u>

11101110

The output is a 1 if any of the inputs is a 1.

Questions

1. Complete the following AND sums:

   (a) 11000011   (b) 11010111

       <u>10011001</u>   <u>01110100</u>

   (c) 10111110   (d) 11000010

       <u>00100100</u>   <u>00111101</u>

2. Complete the following OR sums:

   (a) 10000001   (b) 00001100

       <u>11001100</u>   <u>10100101</u>

Applications of Logic Functions

Masking It is sometimes necessary to SET (to 1) or RESET (to 0) a single bit of an 8-bit byte. This is done by a process called **masking**. For example, if it is necessary to set bit 3 of a byte, then that byte should be ORed with:

00001000 (note that bit 0 is the Least Significant Bit (LSB))

This is known as the **mask byte**. When any two numbers are ORed together, a 1 in any position causes a 1 in the output. This mask byte would cause bit 3 to be set to 1, i.e. if the byte was

```
            11000011:
mask    00001000        OR
result  11001011
```

If a bit needs to be RESET, the AND function is used. ANDing a number with 11110111 would cause bit 3 to be RESET and all others to be left unchanged. This is because 0 ANDed with 1 = 0, 1 ANDed with 1 = 1, while 0 ANDed with 0 = 0 and 1 ANDed with 0 = 0.

For example, to reset bit 7 – AND with 01111111.

```
11000110
01111111        AND
01000110
```

If the supplied byte was 11000110 then the result would be 01000110, i.e. unchanged apart from bit 7.

Complementing

Occasionally, it is necessary to change the state of one or more bits. The simplest way to do this is with the EXCLUSIVE-OR function.

The mask on the lower half of the sum modifies the original byte as shown. When the mask bit is '0' the source bit is unchanged, and when the mask bit is '1', the source bit is complemented.

```
            01010101
EX-OR    00001111
            01011010
```

Questions

1. What mask byte would be used to set

   (a) bits 0 & 7   (b) bit 6

   (c) bits 2 & 4   (d) the lower nibble

2. What mask would be used to clear

    (a) the upper nibble   (b) the lower nibble

3. What mask would be used to change the states of

    (a) bit 4   (b) bits 7 & 1

Arithmetic

Addition

You have probably grasped the practicalities of decimal addition quite well by now. Adding 27 to 31 is easy, but given 89 added to 47, then you have to start using the concepts of 'carries'. The same occurs in binary:

    0   +   1   =   1

    1   +   1   =   0   carry   1

(In decimal 1 + 1 = 2, which is binary 10.)

If a carry is involved, then 1 + 1 + 1 = 1 carry 1 (in decimal, 1 + 1 + 1 = 3 which is binary 11).

Subtraction

Subtraction involves the concept and possibility of negative numbers. An integer byte as used so far does not deal in negative numbers, so the theory must be altered. The number system is altered so that the Most Significant Bit (MSB) – which is bit 7 in a byte – is used to indicate the sign of the number. 10000000 is the smallest number which is –128, while 01111111 is the largest number which is +127.

The process of converting a positive to a negative number is called 'taking its 2's complement'. The process of handling numbers which indicate positive and negative numbers is called '2's complement arithmetic'.

The steps involved are:

1. Change all the 1s to 0s and all the 0s to 1s (this is called the 1's complement).

2. Add 1.

For example, form the 2's complement of→ 01101001   105

    1's complement =   10010110

    add 1   =   00000001

    result   10010111   (–)23

The '–23' refers to the fact that the negative version of '105' is 23 up from the maximum negative number of –128. (Work it out and see!)

To **subtract** 105 from 123, the procedure is to convert 105 into its negative equivalent and then **add** the two numbers.

    105 in binary   01101001

    2's complement   10010111

    123 in binary   01111011

    addition   00010010 = 18

Binary Coded Decimal

Binary Coded Decimal (or BCD) is another binary encoding scheme which encodes each decimal digit into a 4-bit binary code. It is a somewhat inefficient method since only 10 of the possible 16 codes are used. However, it finds a very important use where numeric displays are involved. For example, calculators, tills etc. An 8-bit byte would be encoded as:

    80   40   20   10   8   4   2   1

This could express numbers in the range 00_{10} to 99_{10}.

For example, 56_{10} would be 010101102.

Gray Code

This is commonly used for mechanical shaft-angle encoders, among other things. It has the property that only one bit changes in going from one state to the next. This prevents errors, since with conventional binary, there is no way of guaranteeing that all bits will change simultaneously at the boundary between two encoded values. For example, in going from 7 (0111) to 8 (1000), as a mechanical shaft rotates, it *could* be possible to generate 15 (1111) for a very short period of time.

Gray code avoids this. A 4-bit Gray code would be:

0000	0001	0011	0010
0110	0111	0101	0100
1100	1101	1111	1110
1010	1011	1001	1000

American Standard Code for Information Interchange (ASCII)

Whenever characters, numbers and other codes need to be send from one device to another, it is frequently encoded into ASCII. Originally, this was a 7-bit code, but now has been extended into an 8-bit code to cover a greater variety of symbols. Most of the codes are directly for symbols, but the first 32 are used as control characters. They control things like serial transmission of data, or printers, or file exchange (see page 189).

Table B.1

Ch	hex	dec	ch	hex	dec	ch	hex	dec	ch	hex	dec
NUL	00	00	space	20	32	@	40	64	`	60	96
SOH	01	01	!	21	33	A	41	65	a	61	97
STX	02	02	"	22	34	B	42	66	b	62	98
ETX	03	03	#	23	35	C	43	67	c	63	99
EOT	04	04	$	24	36	D	44	68	d	64	100
ENQ	05	05	%	25	37	E	45	69	e	65	101
ACK	06	06	&	26	38	F	46	70	f	66	102
BEL	07	07	'	27	39	G	47	71	g	67	103
BS	08	08	(	28	40	H	48	72	h	68	104
HT	09	09	)	29	41	I	49	73	i	69	105
LF	0A	10	*	2A	42	J	4A	74	j	6A	106
VT	0B	11	+	2B	43	K	4B	75	k	6B	107
FF	0C	12	,	2C	44	L	4C	76	l	6C	108
CR	0D	13	–	2D	45	M	4D	77	m	6D	109
SO	0E	14	.	2E	46	N	4E	78	n	6E	110
SI	0F	15	/	2F	47	O	4F	79	o	6F	111
DLE	10	16	0	30	48	P	50	80	p	70	112
DC1	11	17	1	31	49	Q	51	81	q	71	113
DC2	12	18	2	32	50	R	52	82	r	72	114
DC3	13	19	3	33	51	S	53	83	s	73	115
DC4	14	20	4	34	52	T	54	84	t	74	116
NAK	15	21	5	35	53	U	55	85	u	75	117
SYN	16	22	6	36	54	V	56	86	v	76	118
ETB	17	23	7	37	55	W	57	87	w	77	119
CAN	18	24	8	38	56	X	58	88	x	78	120
EM	19	25	9	39	57	Y	59	89	y	79	121
SUB	1A	26	:	3A	58	Z	5A	90	z	7A	122
ESC	1B	27	;	3B	59	[	5B	91	{	7B	123
FS	1C	28	<	3C	60	\	5C	92			7C	124
GS	1D	29	=	3D	61	]	5D	93	}	7D	125
RS	1E	30	>	3E	62	^	5E	94	~	7E	126
US	1F	31	?	3F	63	_	5F	95	DEL	7F	127

Index